The Construction of Modern Science:
Mechanisms and Mechanics

Richard S. Westfall

A VOLUME IN THE WILEY HISTORY OF SCIENCE SERIES

The Construction of
Modern Science

HISTORY OF SCIENCE

Editors

GEORGE BASALLA
University of Delaware

WILLIAM COLEMAN
Northwestern University

Biology in the Nineteenth Century:
Problems of Form, Function, and Transformation

WILLIAM COLEMAN

Physical Science in the Middle Ages

EDWARD GRANT

The Construction of Modern Science:
Mechanisms and Mechanics

RICHARD S. WESTFALL

The Construction of Modern Science

Mechanisms and Mechanics

RICHARD S. WESTFALL

Indiana University

John Wiley & Sons, Inc.

New York · London · Sydney · Toronto

To Alfred, Jennifer, and Kristin

Series Preface

THE SCIENCES CLAIM an increasingly large share of the intellectual effort of the Western world. Whether pursued for their own sake, in conjunction with religious or philosophical ambitions, or in hopes of technological innovation and new bases for economic enterprise, the sciences have created distinctive conceptual principles, articulated standards for professional training and practice, and have brought into being characteristic social organization and institutions for research. The history of the sciences—astronomy; physics and associated mathematical methods; chemistry; geology; biology and various aspects of medicine and the study of man—consequently exhibits both great interest and exceptional complexity and presents numerous difficulties for investigation and interpretation.

For over half a century an international group of scholars have been studying the historical development of the sciences. Such studies have often called for an advanced level of scientific competence on the part of the reader. Furthermore, these scholars have tended to write for a small specialist audience within the history of science. Thus it is paradoxical that the ideas of men who are professionally committed to elucidating the conceptual development and social impact of science are not readily available to the modern educated man who is concerned about science and technology and their place in his life and culture.

The editors and authors of the *Wiley History of Science Series* are dedicated to bringing the history of science to a wider audience. The books comprising the series are written by men who are fully familiar with the scholarly literature of their subject. Their task, and it is not an easy one, is to synthesize the discoveries and conclusions of recent scholarship in history of science and present the general reader with an accurate, short narrative and analysis of the scientific activity of major periods in Western history. While each volume is complete in itself, the several volumes taken together will offer a comprehensive general view of the Western scientific tradition. Each volume, furthermore, includes an extensive critical bibliography of materials pertaining to its topics.

George Basalla
William Coleman

Preface

FOR THE PAST SEVEN YEARS I have been teaching the history of
science in the 17th century. This textbook, addressed to the average
undergraduate, gives a summary statement of my understanding of the
subject. I realize that my understanding has not reached (or even approached) a final configuration; and I suspect that if I were to rewrite
this book five years from now I would devote more space to Renaissance
Naturalism (or the Hermetic tradition, as I sometimes call it) and the
sociological forms in which the scientific movement clothed itself.
Nevertheless, I do not think that the changes would wholly transform
the present volume. Instead, they would constitute modifications of a
structure that aspires to present a coherent interpretation of the scientific
revolution that will have more than ephemeral value.

Inevitably I have acquired numerous obligations. I am grateful to
Indiana University and its Department of History and Philosophy of
Science for the opportunity to devote myself to the continued study
that was necessary to write this book. I thank various libraries, especially
those of Cambridge University, Harvard University, and Indiana University for the use of their facilities and service. My students gave me
the chance to test ideas against their beneficient skepticism. My col-
leagues at Indiana University and elsewhere rendered informed counsel
and criticism. My family gave me constant support without which
none of the opportunities could have had any effect. "And to be specific
at last, to my son Alfred I owe the Index."

Richard S. Westfall

Acknowledgments

THE DIAGRAMS ILLUSTRATING the devices of Ptolemaic astronomy are reprinted by permission of the publishers from Thomas S. Kuhn, *The Copernican Revolution,* Cambridge, Mass.: Harvard University Press, copyright 1957 by the President and Fellows of Harvard College. The quotations from Galileo, *Dialogue Concerning the Two Chief World Systems,* trans. Stillman Drake (Berkeley, 1962); William Harvey, *Lectures on the Whole of Anatomy* (Berkeley, 1961); and Isaac Newton, *Mathematical Principles of Natural Philosophy,* trans. Florian Cajori (Berkeley, 1960), were all originally published by the University of California Press and are reprinted by permission of The Regents of the University of California. The passage from Giambattista della Porta's *De refractione* is quoted by the permission of Professor Vasco Ronchi from his *Storia della luce* (Bologna, 1939). The diagram illustrating Galenic physiology and the passages from Thomas Moffett and Francesco Stelluti are taken from Charles Singer, *A History of Biology* copyright 1931, 1950, and 1959 by Abelard-Schuman, Ltd., by permission of the publisher. The diagram of the structure of the heart, from Spalteholz-Spanner, *Atlas of Human Anatomy,* 10th ed., trans. A. Nederveen (Philadelphia, 1967), is included by permission of the publisher, F. A. Davis Company. The quotations from Malpighi and Swammerdam are reprinted from Howard B. Adelmann, *Marcello Malpighi and the Evolution of Embryology,* copyright 1966 by Howard B. Adelmann, used by permission of Cornell University Press. Martinus Nijhoff has permitted me to use diagrams for the analysis of the physical pendulum and the cycloid from the *Oeuvres complètes* of Christiaan Huygens (La Haye, 1888–1950). Quotations from the dynamical essays of Leibniz are reprinted with the permission of D.

Reidel Publishing Company from Leibniz, *Philosophical Papers and Letters,* Leroy Loemker, Ed. (Chicago, 1956). The Syndics of the Cambridge University Library have granted permission to quote from Isaac Newton's manuscript "Waste Book" and to reproduce diagrams from it.

Contents

The Construction of
Modern Science

Introduction

TWO MAJOR THEMES dominated the scientific revolution of the 17th century—the Platonic-Pythagorean tradition, which looked on nature in geometric terms, convinced that the cosmos was constructed according to the principles of mathematical order, and the mechanical philosophy, which conceived of nature as a huge machine and sought to explain the hidden mechanisms behind phenomena. This book explores the founding of modern science under the combined influence of the two dominant trends. The two did not always mesh harmoniously. The Pythagorean tradition approached phenomena in terms of order and was satisfied to discover an exact mathematical description, which it understood as an expression of the ultimate structure of the universe. The mechanical philosophy, in contrast, concerned itself with the causation of individual phenomena. The Cartesians at least were committed to the proposition that nature is transparent to human reason, and mechanical philosophers in general endeavored to eliminate every vestige of obscurity from natural philosophy and to demonstrate that natural phenomena are caused by invisible mechanisms entirely similar to the mechanisms familiar in everyday life. Pursuing different goals, the two movements of thought tended to conflict with each other, and more than the obviously mathematical sciences were affected. Since they proposed conflicting ideals of science and differing methods of procedure, sciences as far removed from the Pythagorean tradition of geometrization as chemistry and the life sciences were influenced by the conflict. The explication of mechanical causation frequently stood athwart the path that led toward exact description, and the full fruition of the scientific revolution required a resolution of the tension between the two dominant trends.

The scientific revolution was more than a reconstruction of the categories of thought about nature. It was a sociological phenomenon as well, both expressing the ever increasing numbers engaged in the activity of scientific research and spawning a new set of institutions that have played a more and more influential role in modern life. In my opinion, however, the development of ideas following their own internal logic was the central element in the foundation of modern science and, although I have attempted to indicate something of the sociological ramifications of the scientific movement, this book expresses my conviction that the history of the scientific revolution must concentrate first of all on the history of ideas.

CHAPTER I

Celestial Dynamics and Terrestrial Mechanics

WHEN THE 17TH CENTURY dawned, the Copernican revolution in astronomy was over fifty years old. Perhaps one should say rather that Copernicus' book, *De revolutionibus orbium coelestium** (1543), was over fifty years old. Whether the book would initiate a revolution had yet to be determined, and two men who had scarcely passed the thresholds of their scientific careers in 1600 were to be the primary agents in assuring that it would. Both Johannes Kepler (1571–1630) and Galileo Galilei (1564–1642) acknowledged Copernicus as their master; both devoted their careers to confirming the revolution in astronomical theory he had begun. To its confirmation each made an essential contribution, though in his contribution each modified Copernicanism in a way the master might not have accepted. Copernicus himself had proposed a limited reformation of planetary theory within the broad outlines of the accepted framework of Aristotelian science. By the time Kepler and Galileo were done, the limited reformation had become a radical revolution, and the work of the 17th century, which laid the foundation for the structure of modern science, consisted in pursuing the questions that Kepler and Galileo opened. Intellectual history does not always divide neatly into packages that fit the calendar, and scientists have not concerned themselves to group their labors into units convenient to the academic curriculum. The dawn of the 17th century, however, did coincide with the dawn of a new era in science.

Kepler had made his professional debut four years earlier with the publication of *Mysterium Cosmographicum** in 1596. To 20th century

* *On the Revolutions of the Heavenly Spheres.*
* *Cosmographic Mystery.*

eyes, the book appears even more mysterious than the title promises; but when it is probed, its mystery illuminates much of Kepler's work. Avowedly Copernican, the book set out to demonstrate the validity of the heliocentric theory from the number of planets. Because the moon was considered a planet in the Ptolemaic system, the Copernican system had one less planet, six instead of seven. Kepler undertook to demonstrate why God had chosen to create a universe with six planets, that is, a heliocentric universe. God's choice, as it turned out, had been dictated by the existence of five, and only five, regular solids. If a cube were inscribed inside the sphere defined by the radius of Saturn, the radius of the sphere inscribed inside the cube would be that of Jupiter, and so on. The five regular solids define the spaces between six spheres, and because only five regular solids exist, only six planets exist. The question *Mysterium Cosmographicum* asked is not the sort that modern science has tended to pose. Just for that reason, it reveals more clearly the fundamental assumptions with which Kepler approached his work in astronomy. Like Copernicus before him, Kepler had drunk deeply at the spring of Renaissance neoplatonism, and imbibed its principle that the universe is constructed according to geometric principles. Coming two generations later, Kepler had the perspective to see where Copernicus' system failed to achieve the ideal of geometrical simplicity which both of them shared. Kepler's work would be the perfection of Copernican astronomy according to neoplatonic principles.

Kepler was equally convinced that astronomical theory must be more than a set of mathematical devices to account for observed phenomena. It must rest on sound physical principles as well, deriving the motions of planets from the causes producing them. To his greatest work he gave the title *New Astronomy Founded on Causes, or Celestial Physics Expounded in a Commentary on the Movements of Mars.** Since the time of Aristotle, nearly two thousand years before Kepler, there had been virtual unanimity that, physically speaking, the heavens were constructed of crystalline spheres. The perfection and immutability ascribed to the celestial realm required a material different from the four elements that composed the corruptible bodies of the mundane world, and the axial rotation of the spheres, the one movement allowed to the heavens, corresponded to the perfect circular motion from which astronomers were expected to construct their theories. The "Celestial Spheres"

---

* In the original Latin (and Greek): *Astronomia nova* ΑΙΤΙΟΛΟΓΗΤΟΣ *seu physica coelestis tradita commentariis de motibus stellae Martis.*

referred to in Copernicus' title were the same crystalline spheres. Kepler, however, was convinced that crystalline spheres do not exist. Careful observations by Tycho Brahe and others of the new star of 1572 and of the comet of 1577 had demonstrated that both were located in the realm beyond the moon, which was claimed to be immutable. The motion of the comet appeared to be incompatible with the existence of crystalline spheres. "There are no solid spheres as Tycho Brahe has demonstrated"—the phrase runs like a refrain through Kepler's works. And if the crystalline spheres had been shattered, a new celestial physics must be established to account for the stable, recurring motions of the planets. The constant search for physical causes went hand in hand with the search for geometrical structure—to Kepler, the two were only different aspects of a single reality.

The physical principles he employed expressed the basic propositions of Aristotelian dynamics, and the 17th century replaced them with a wholly different set. Nevertheless, Kepler was the founder of modern celestial mechanics. He was the first to insist categorically that the long-accepted crystalline structure of the heavens did not exist and that a new set of questions about celestial motions had to be formulated. Convinced of the uniformity of nature, he attempted to account for the phenomena by the same principles employed in terrestrial mechanics. More than anything else, this aspect of Kepler's thought makes him a revealing figure in the early history of modern science. In him we can observe a celestial mechanics, founded on the principles of terrestrial mechanics, begin to replace the purely kinematic treatment of the heavens. An astronomy that sought to comprehend the forces controlling planetary motions supplanted the manipulation of circles that were deemed to express the perfection and incorruptibility of a realm apart. If Kepler's dynamic principles ultimately revealed themselves to be unsatisfactory, he followed them, nevertheless, to the laws of planetary motion that are accepted today.

It was, of course, the real mathematical structure and the real physical causes which Kepler sought to uncover. Such must square with the observations, and Kepler refused to force a priori theories onto nature in violence of the observed facts. Here lay the problem of the *Mysterium Cosmographicum*. In the cases of Mercury and Saturn, the theory diverged widely from the accepted observations. Kepler was aware, however, that the accepted observations were unreliable, and that a contemporary observer, Tycho Brahe, was collecting a body of data more accurate by far. In 1600, Kepler became Tycho's assistant. In 1601,

Tycho died; and with no right whatever beyond the prerogative of genius, Kepler simply appropriated the precious body of observations. They served as the irreplaceable data on which his genius worked to develop the laws of planetary motion.

Mars was to be the principal object of his labor. Kepler, who always asserted the structural unity of the solar system, would not hesitate to apply his conclusions on Mars to the other planets. The *Astronomia Nova,* published in 1609, embodied the conclusions. But it contained much more as well. An intellectual autobiography, it described in detail every step of the investigation, so that we can follow the progress of Kepler's thought in a way that is possible with few other scientists. The progression of thought revealed was twofold—on the one hand, there was a movement away from the age-old obsession with circularity and toward the acceptance of noncircular orbits; on the other hand, there was a movement away from animistic modes of thought and toward a frankly mechanistic conception of the universe.

Ever since the flowering of Greek science, astronomy had attempted to account for celestial phenomena by combinations of uniform circular motions. The circle being the perfect figure, it alone was suitable to describe the heavens. Kepler too began his consideration of Mars with a circle, but from the beginning his treatment differed from earlier ones. Astronomers before him had combined circles—using a basic deferent, as it was called, with whatever combination of eccentrics and epicycles an individual might choose—to account for observed positions of the planets. (See Fig. 1.1.) The vectorial addition of the radii, laying them end to end, must place the planet where observations found it to be. In contrast, Kepler, who was convinced that new physical considerations must prevail, that crystalline spheres do not exist, but that planets nevertheless follow definite orbits through the immensity of space, was concerned from the beginning with the orbit itself. No previous theory had proposed that the path of a planet is a circle. Kepler first attempted to fit Mars to just such a circular orbit. Even in utilizing the circle, however, Kepler began to reject it, by denying uniform circular motion and accepting, as the evidence demanded, the proposition that Mars moves in its orbit with a varying velocity.

After Kepler had invested two years of effort in the theory, it finally failed. It contained an inaccuracy of 8′. Copernicus before him had been satisfied with an accuracy of 10′; Kepler could not forget, however, that Tycho's observations imposed a higher standard. "Since divine goodness has granted us a most diligent observer, Tycho Brahe, from whose ob-

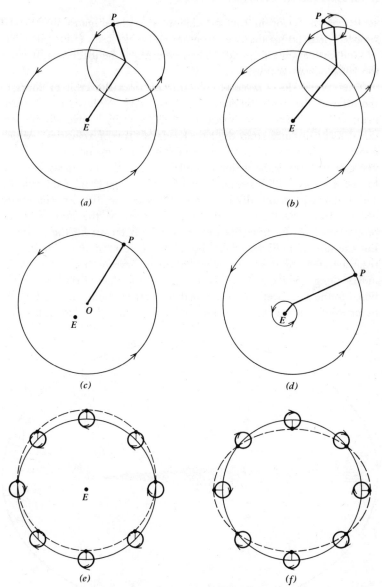

Figure 1.1. The geometrical devices of Ptolemaic astronomy. (a) A major epicycle on a deferent. (b) An epicycle on a major epicycle. (c) An eccentric. (d) An eccentric on a deferent. (e) The effect of a minor epicycle with the same period as the deferent. (f) The effect of a minor epicycle with a period twice that of the deferent.

servations the error in this calculation of eight minutes in Mars is revealed, it is fitting that we recognize and make use of this good gift of God with a grateful mind." The first use of it he made was to reject the labor of two years.

Temporarily discouraged, Kepler turned from the orbit of Mars to the orbit of the earth. Extending the principles employed in his treatment of Mars, he concluded that the earth's velocity is inversely proportional to its distance from the sun. Kepler's "law of velocities," which Newton proved to be incorrect, served as a guiding beacon to his investigation. From it, he deduced the law of areas, which today we hold to be correct and call his second law of planetary motion. If the velocity varies inversely as the distance from the sun, the distance (or radius vector) from the sun of every small segment of the orbit must be proportional to the time the planet spends in traversing the segment. But the sum of radii vectors to the small segments of the orbit may be regarded as equal to the area that the radius sweeps out as the planet moves along. (See Fig. 1.2.) That is to say, the elapsed time is proportional to the area swept out. The mathematical reasoning was fallacious; never mind, the law of velocities used as a premise was also fallacious,

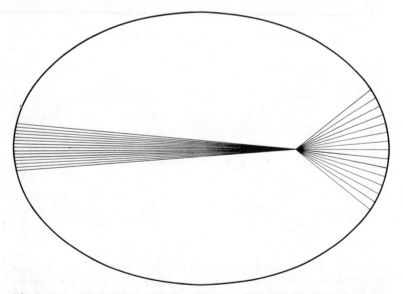

Figure 1.2. Kepler's law of areas. The eccentricity of the ellipse has been greatly exaggerated. The space between each pair of lines represents a single unit of time.

but the conclusion has proved to be correct. The law of areas served a specific technical need. In the old astronomy of deferents and epicycles, the position of a planet could be calculated by the vectorial addition of radii, each of which turned at a uniform rate. Much of the power of the circle in astronomy consisted in its technical utility. Having abolished the machinery of multiple circles in favor of a single circle on which a planet moves with a nonuniform velocity, Kepler needed a formula by which to calculate the planet's position. This the law of areas supplied. And in supplying it, the law of areas made the circle dispensable in astronomy as it had never been before.

Kepler had derived the law of areas from the (erroneous) law of velocities. The law of velocities also suggested the basic elements of his celestial mechanics, which depended on the central dynamic function assigned to the sun. Kepler was convinced of the primary role of the sun in the universe. The source of all light and all heat, the sun must also be the source of all movement, the dynamic center of the solar system. Kepler imagined some power to radiate out from the sun, like the spokes of a wheel. As the sun turned on its axis the spokes would push the planets along. (See Fig. 1.3.) Nothing in Kepler's celestial mechanics operated to pull a planet aside from a tangential path and retain it in an orbit around the sun. The continuing hold of the circle over the thought even of the man who broke its grip on astronomy is attested by the fact that Kepler never doubted that planets would move round the sun in closed orbits if they moved at all. Obviously Kepler was employing the basic propositions of Aristotelian mechanics, according to which a body remains in motion only as long as something continues to move it, its velocity being proportional to the moving force. Thus the law of velocities appeared as an obvious consequence of the basic dynamics of the solar system. The effectiveness of the power radiating from the sun should decrease in proportion to the distance, and the velocity of each planet should vary inversely as its distance from the sun.

The more Kepler contemplated the dynamics of planetary motion, the more it recalled the basic relations of the lever. The farther a planet was removed from the sun, the less the power of the sun was able to move it. When the concept of a power radiating out from the sun first appeared in the *Mysterium Cosmographicum*, Kepler called it an *"anima motrix,"* a "motive soul," a phrase redolent of animistic connotations. In 1621, as he prepared a second edition of the *Mysterium*, he added a footnote: "If you substitute the word 'force' [*vis*] for the

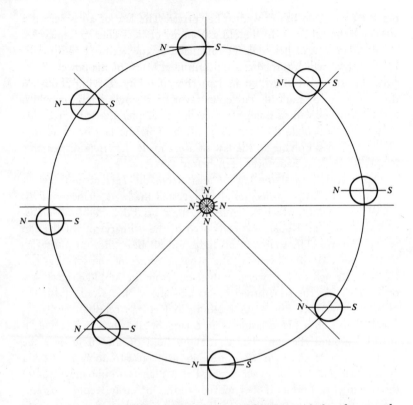

Figure 1.3. Kepler's celestial mechanics. As the planet circles the sun, the position of its axis maintains a constant alignment. The sun is a peculiar magnet, its surface constituting one pole and its center the other. Through half its orbit the planet is attracted toward the sun; through the other half it is repelled.

word 'soul' [*anima*], you have the very principle on which the celestial physics in the *Commentary on Mars* [*Astronomia Nova*] is based. For I formerly believed completely that the cause moving the planets is a soul, having indeed been imbued with the teaching of J. C. Scaliger on motive intelligences. But when I recognized that this motive cause grows weaker as the distance from the sun increases, just as the light of the sun is attenuated, I concluded that this force must be as it were corporeal." From *anima motrix* to *vis,* from the animistic to the mechanistic —the development in Kepler's thought foreshadowed the course of 17th century science.

Meanwhile one problem in his celestial dynamics remained to be solved. What causes a planet's distance from the sun to vary? Kepler's pursuit of the issue led him ever further from the circular orbit. The astronomical tradition presented an obvious answer to the variation of distance—an epicycle turning on the basic deferent. It is testimony to the power that the tradition of circles exercised over Kepler that he attempted initially to explain the variation by an epicycle. An epicyclic mechanism affronted his sense of physical reality, however. A planet would need intelligence to turn on an epicycle around a moving point not occupied by a body. When he returned to the consideration of Mars, he discovered that when he used an ellipse to approximate the orbit, which he now assumed to be oval, the radius vector varied in length according to a uniform sine function. The uniform variation suggested a purely physical action which required no supervisory intelligence. The mechanism of epicycles could be rejected at last, once and for all. He perceived it, Kepler said, "as one aroused from sleep who gazed with astonishment on a new light." Kepler ultimately decided that a magnetic action from the sun attracts a planet during half of its orbit while one pole is presented to the sun and repels it during the second half while the other pole is presented. (See Fig. 1.3.) Meanwhile the hold of the circle had been broken, and Kepler went on to conclude that the orbit does more than approximate an ellipse. It is an ellipse at one focus of which the sun is located—a conclusion we call his first law of planetary motion.

Although Kepler later discovered what is called his third law (relating the period (T) of each planet to its mean radius (R), so that $T^2/R^3 =$ a constant, for the solar system), the immediate importance of his work lay in the first two. Nearly a century earlier, Copernicus had set out to find a planetary system that would satisfy the demand for geometric simplicity. Kepler solved Copernicus' problem, carrying simplicity to a level not even dreamed of before in the history of astronomy. If Copernicus' initial assumption, that the sun instead of the earth is the center of the solar system, were granted, one conic curve sufficed to describe the orbit of each planet. All of the complexity of eccentrics and epicycles had been swallowed up in the simplicity of the ellipse. The bait concealed a hook, of course. The cost of accepting the ellipse's simplicity was the abandonment of the circle, with all its ancient connotations of perfection, immutability, and order. Only by degrees and then only imperfectly had Kepler freed himself from the circle's power over his imagination, and he never forgot what its attractions were. The chief value of the second law, in his eyes, was the new uniformity it offered to replace that of circular motion. To a friend who protested against the ellipse, he described the circle as a voluptuous whore enticing astronomers away from the honest maiden nature. His master, Copernicus, had preferred the jade. If it is true to say that Kepler perfected Copernican astronomy, it is equally true to say that he destroyed it.

At least half the fascination and the perplexity of Kepler lies in the fact that what we call his three laws are hidden under a mountain of speculation that could hardly be more foreign to the mentality of the 20th century—speculation relating musical harmonies to planetary motions, speculation on the geometrical architecture of the universe, and, not least, speculation on celestial dynamics employing conceptions soon to be replaced. How are we to explain the derivation of laws we accept as accurate from principles we have long since rejected? To explain the anomaly we must distinguish the means of discovery from the means of verification. Kepler's laws have stood the test of time because they conform to observed facts. In Tycho's body of data he had a set of reliable observations, and he refused to accept a conclusion that contradicted them. How was he to proceed toward any conclusion? All that the observations gave were the positions of the planets among the fixed stars—lines along which planets were located at the times of observation. To imagine that Kepler could simply plot them on a diagram and recognize the resultant ellipse is to suggest the impossible. If it had been possible, astronomy would not have waited for Kepler to discover elliptical orbits. Principles to guide the investigation were needed, and all of the old principles appeared to be crumbling. What a world of implication was involved in his assertion that the crystalline sphere had been destroyed. The very structure of the universe, long accepted as beyond question, had been called into doubt and rejected. Kepler's principles provided the bases without which there could have been no investigation at all, and however strange we may find them, we must not ignore the role they played. If a new science of mechanics was soon to replace his physical principles, let us not forget that Kepler first drew the full implications of the new situation in astronomy and posed the question of celestial dynamics. Posing a question correctly, in science as in other areas of study, is more important than giving the answer, and science has treated celestial motions as problems in mechanics ever since.

What should the reaction of an ordinary intelligent man have been to Kepler's version of heliocentric astronomy? Its advantages as a geometric hypothesis were obvious, but was there any reason to accept it as the true system of the universe? When the reasons are examined, it appears that its advantage as an hypothesis was the principal reason for accepting it. That is, aside from its geometrical simplicity there was very little evidence in its favor. True, the telescope had been invented, and in 1609 Galileo had turned it on the heavens. He had observed a number of things that tended to support a heliocentric system, but nearly all of them merely strengthened arguments already advanced on other grounds. The craters on the moon and the spots on the sun appeared to contradict the perfection and immutability of the heavens, but the new star of 1572 and the comet of 1577 had already done as much. The satellites of Jupiter were another matter perhaps. Before their discovery, the moon, a planet circling a planet as it were, had appeared to be an unexplained anomaly in the heliocentric system and therefore an objection to it. If the satellites of Jupiter did not explain the phenomenon, at least they destroyed its uniqueness, and the moon appeared to be less anomalous. The satellites of Jupiter offered no positive support to the heliocentric system, however. The phases of Venus did. In the geocentric system, Venus is always more or less between the sun and the earth, and must always appear as a crescent. In the heliocentric system, it travels behind the sun and can appear nearly full—which the telescope revealed, of course. (See Fig. 1.4.) There was one other thing which the telescope did not reveal, however, and as far as the Copernican revolution is concerned, it was the most perplexing telescopic observation. The telescope did not reveal stellar parallax. From the moment when the Copernican system was born, the crucial relevance of stellar parallax had been obvious. If the earth travels around the sun on an immense orbit, the positions of the fixed stars should change as an observer moves from one end of the orbit to another. (See Fig. 1.5.) No stellar parallax, as it is called, appeared to the naked eye. None appeared through the telescope either. As we know today, the fixed stars are so far removed that telescopes of considerable power, not developed until the 19th century, are required to distinguish the very small angle. Galileo's telescope could not distinguish it, and the nonappearance of stellar parallax balanced, at the very least, the positive evidence offered by the phases of Venus. The case for the Copernican-Keplerian system stood or fell on the argument of geometric harmony and simplicity. For this advantage and for little else, men were asked to overturn a con-

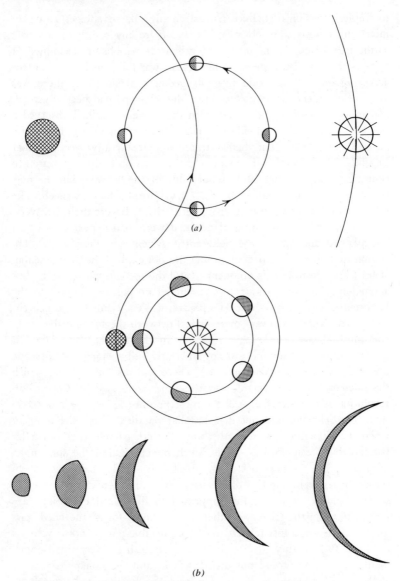

Figure 1.4. *Phases of Venus. (a) Ptolemaic system. (b) Copernican system. In the Ptolemaic system, Venus must always appear more or less crescent shaped. In the Copernican system, it can appear nearly full as it passes behind the sun, and its size varies greatly.*

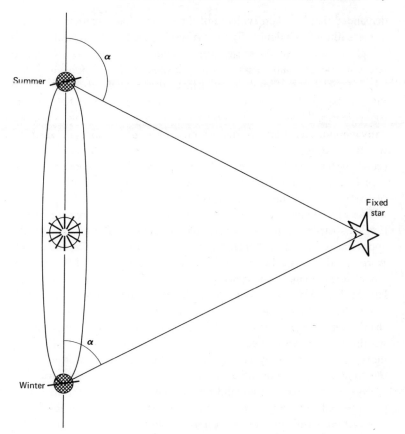

Figure 1.5. Stellar parallax. The earth's orbit is shown from the side. For positions of the earth six months removed from each other, the two angles α at which a fixed star is observed ought to differ from each other if the earth is in fact traveling around the sun.

ception of the universe which included as well physical, philosophical, psychological, and religious questions of the most all-embracing nature. Perhaps it was more of a load than geometric simplicity could bear.

Not the least of the sacrifices demanded in the name of simplicity was common sense itself. It has been remarked many times that modern science has required a re-education of common sense. What could have been more common-sensical than a geocentric universe? We still say that the sun rises and speak of solid earth. The heliocentric universe demanded that the plain evidence of the senses in such matters be denied as mere illusion. Undoubtedly the major obstacle to the new astronomy's acceptance was common sense, which ridiculed it every day. Common sense, moreover, found a sophisticated expression in the prevailing doctrine of motion. As Simplicius said in Galileo's *Dialogue,* "The crucial thing is being able to move the earth without causing a thousand inconveniences." The inconveniences referred to were primarily concerned with motion. According to the accepted ideas of motion, the assertion that the earth is turning daily upon its axis was absurd. Before the heliocentric system could be generally accepted, the inconveniences had to be explained away, and the man who did so was the man who put the phrase into Simplicius' mouth, Galileo Galilei.

From its beginning, Galileo's career focused upon the science of motion. His earliest important work, stemming from the early 1590's, the approximate period of Kepler's first work, was entitled *De motu.** *De motu* reveals that Galileo began his career as an adherent of the impetus school of mechanics. The concept of impetus had developed during the late Middle Ages as a solution to the problem on which Aristotle's mechanics stumbled most conspicuously. Aristotle had based his mechanics on the principle—in itself as obvious to common sense as the stability of the earth—that every motion requires a cause, that a body moves as long, and only as long, as something moves it. Considered in the light of the motion of a cart pulled by an ox, or even that of a galley pulled by oars (if one did not examine it too closely), the principle appeared so obvious as to be trite. Greeks had also thrown the discus, however, and with projectiles such as the discus difficulties arose. What keeps a projectile in motion once it has separated from the projector? Aristotle answered by referring the cause to the medium through which the projectile moves. The concept of impetus, on the other hand, transferred the cause of the continuing motion—the necessary cause, required by the nature of motion—from the medium to the projectile. A body put in motion acquires an impetus which continues to move it after its separation from the projector. From the fourteenth through the sixteenth century, the concept of impetus had stood in the vanguard of creative thought on mechanics, and it is not surprising that Galileo embraced it in his youth. To the concept of impetus he joined the influence of Archimedes, finding a way to interpret impetus in terms of fluid statics, and attempting by this means to construct an exact quantitative dynamics

* *On Motion.*

to supplement Archimedes' statics. Although he repudiated the concept of impetus within a decade, *De motu* set the tone of Galileo's scientific work. Throughout his career he pursued the ideal of a quantitative science of motion, and the scientific revolution built its proudest achievement, its mechanics, on the foundation he provided.

Galileo abandoned the mechanics of *De motu* when it proved itself unable to solve the basic problem to which he addressed himself. That problem was the apparent contradiction between the phenomena of motion we observe about us and the assertion that the earth rotates daily on its axis. Suppose that a ball is dropped from a tower. According to the Copernican system, the tower is travelling at an immense speed from west to east. As soon as the ball is released and the force of the hand which has been impelling it to move with the tower ceases to operate, its eastward motion should halt; and as it falls to the earth with the natural motion of a heavy body, it should appear to fall well to the west of the tower. In fact, of course, we all know that a ball falls straight down, parallel to the tower's side. Hence the earth cannot possibly be turning on its axis. While the inconveniences entailed by a moving earth, which Simplicius insisted on in Galileo's *Dialogue*, could be expressed in many ways, the problem of vertical fall can stand as a reasonable epitome of them all. One must understand that the objection was not ridiculous. According to Aristotle's conception of motion, that is, according to the system of mechanics accepted by everyone, it was absurd to suggest that the earth is in motion. To be answered, the objection required the creation of a new system of mechanics.

Briefly stated, the solution to the problem set by Copernican astronomy, and the foundation of the new mechanics, was the concept of inertia. A body in motion continues to move with a uniform velocity until something external operates to change it. "Keeping up with the earth," Galileo stated in response to the problem of the falling ball, "is the primordial and eternal motion ineradicably and inseparably participated in by this ball as a terrestrial object, which it has by its nature and will possess forever." Since no cause operates to check the motion of the ball from west to east, it keeps up with the tower from which it is dropped as it falls to the earth. In one of his Socratic exchanges with Simplicius, Salviati (Galileo's mouthpiece in his great polemic for the Copernican system, *Dialogo sopra i due massimi sistemi del mondo,* * 1632) asks what would happen if a ball were placed on a sloping plane.

* *Dialogue Concerning the Two Chief World Systems.*

It would roll down the plane with a steadily increasing velocity. Would it roll up the plane? Not unless it were given an original impulse and then with its motion constantly retarded. What if it were placed on a horizontal plane and given a push in some direction? There would be, Simplicius agrees, no cause for acceleration or deceleration, and the ball would therefore continue as far as the surface extends. "Then if such a space were unbounded, the motion on it would likewise be boundless? That is, perpetual?" "It seems so to me," the thoroughly worsted Aristotelian replies. As Descartes was later to summarize the issue, men had been asking the wrong question about motion. They had been asking what keeps a body in motion. The proper question was what causes it ever to stop.

Galileo did not employ the word "inertia." For that matter, whatever his phraseology, he did not employ the concept of inertia in precisely the form we hold it today. No man is able wholly to break with the past—not even a giant such as Galileo—and even in formulating a new conception of motion, he was bound by elements of the old cosmology. His universe was not an impersonal universe of mechanical laws and matter in motion. It was a cosmos rather, organized by infinite intelligence. As such, it was ordered, inevitably, according to the perfect figure, the circle. Following the old tradition, Galileo held that circular motion, and circular motion alone, is compatible with an ordered cosmos. Only in a circle can a body move forever in its natural place, maintaining always the same distance from the same point, and only in circular motions can the bodies of the cosmos retain forever their primordial relations. Rectilinear motion implies disorder; a body removed from its natural place returns to it along a straight line. Once there, it remains in its place by resuming a natural circular motion.

Thus the astronomy of the *Dialogue* was such as no professional astronomer could have accepted. Published over twenty years after Kepler's *Astronomia Nova,* the *Dialogue,* which intended to support the heliocentric system, ignored Kepler's conclusions as it ignored the technical necessity for the epicycles of earlier theories. It discussed the Copernican system as though each planet moved in a simple circular orbit. The relation of Galileo to Kepler is fraught with irony. Kepler, who treated the solar system in mechanical terms and sought to comprehend the physical forces that govern its motions, employed a system of mechanics based on principles overthrown by Galileo. Galileo, who formulated the basic concepts of the new mechanics, ignored the problems to which Kepler's celestial mechanics addressed itself and held that the planets move naturally in circular orbits.

Galileo was thinking in similar terms when he confronted the problem of motion on a spinning earth, and the concept of inertia that he formulated reflects the terms in which the problem presented itself to him. As we have seen above, Salviati leads Simplicius to agree that a ball rolling on a horizontal plane does not experience any cause either to accelerate or to slow down and hence should continue forever. What is a horizontal plane? It is of course a plane which is everywhere "equally distant from the center." Inertial motion was conceived as uniform circular motion, the natural motion of a body in its natural place in a well-ordered universe.

Behind the principle of inertia lay a radically new conception of motion itself. To Aristotle, motion had been a process involving the very essence of a body, a process whereby its being was enhanced and fulfilled. Local motion—what alone the word "motion" means to us—had been to Aristotle only one example of a much broader conception meant to embrace all change. The education of a youth or the growth of a plant were motions quite as much as the vertical fall of a heavy body; if anything, they were better examples of the process he had in mind. Just as the seed realizes its full potential by growing into a plant, so a heavy body realizes its nature by moving toward its natural place. The heart of Galileo's conception of motion lay in the separation of motion from the essential nature of bodies. Nothing in a body is affected by its (uniform horizontal) motion. Motion is merely a state in which a body finds itself; and as Galileo repeated over and over, a body is indifferent to its state of motion or rest. Rest indeed is not distinct from motion at all; it is merely "an infinite degree of slowness." The idea of indifference was basic to Galileo's solution to the problem of motion in the Copernican universe. Because we are indifferent to motion, we can be moving with an immense velocity and not perceive it—an assertion which was absurd within an Aristotelian context, in which motion expresses the nature of a body.

"For consider [Galileo argued]: Motion, in so far as it is and acts as motion, to that extent exists relatively to things that lack it; and among things which all share equally in any motion, it does not act, and is as if it did not exist. Thus the goods with which a ship is laden leaving Venice, pass by Corfu, by Crete, by Cyprus and go to Aleppo. Venice, Corfu, Crete, etc. stand still and do not move with the ship; but as to the sacks, boxes, and bundles with which the boat is laden and with respect to the ship itself, the motion from Venice to Syria is as nothing, and in no way alters their relation among themselves. This is so because it is common to all of them and all share equally in it. If, from the cargo in the ship, a sack were shifted from a chest one single inch, this alone would be more of a movement for it than the two-thousand-mile journey made by all of them together."

Motion understood in this sense does not require a cause any more than rest requires one. Only changes of motion require a cause.

Because of a body's indifference to motion, it can participate at the same time in more than one. None of them impedes the others, and they compound together smoothly to trace a trajectory however complex. One of Galileo's supreme achievements was to demonstrate that the horizontal motion of a projectile compounds with its uniformly accelerated fall toward the earth with the result that the body follows a parabolic path. A body is indifferent even to a violent motion such as that of a cannon ball. The most sophisticated exposition of the objections to the earth's motion rested its case on exactly the contention that a body cannot be indifferent to such a motion. Tycho Brahe argued that the extreme violence of a cannon shot cannot fail to obstruct the natural motions said to belong to the ball, and until the violent motion has been exhausted, the natural ones cannot assert themselves. Hence the earth must turn under a cannon ball in midair, and one shot to the west should fall much farther away than one shot to the east. In agreement with a long tradition, Tycho implied that the trajectory of a cannon ball is rectilinear until the violence of the shot is wholly or nearly exhausted. In contrast, Galileo asserted that the trajectory curves from the moment the ball leaves the cannon's mouth. If the idea of natural motion continued to have a place in his thought, the distinction in which it had earlier participated, that between natural and violent motion, had lost its meaning. All motion, as motion, is identical. The same reasoning which explained why a ball drops at the foot of a tower on a turning earth explained as well why it drops at the foot of the mast on a moving galley. Bodies are indifferent to motion—to all motion.

Galileo's conception of inertia, with the further assertion that inertial motion is rectilinear, became the cornerstone for the entire structure of modern physics. As such, it is instilled into all of us in the educational process to the point that we regard it as natural and self-evident. We cannot even examine it objectively, let alone imagine the difficulties of formulating the idea initially in a world predisposed to regard it, not as self-evident, but as self-evidently absurd. Does not the principle of inertia merely express the observed facts of motion? The suggestion embodies our conviction that modern science rests on the solid foundation of empirical fact, that it was born when men turned from the empty sophistries of medieval Scholasticism to the direct observation of nature. Galileo, alas, is difficult to fit into such a picture, and the concept of inertia even more so. Throughout the *Dialogue*, it is Simplicius, Galileo's own creation to expound the viewpoint of Aristotelianism, who asserts the sanctity of observation. Salviati, who speaks for Galileo, has ever to deny the claims of sense in favor of reason's superior right.

"Nor can I ever sufficiently admire the outstanding acumen of those who have taken hold of this opinion [Copernicanism] and accepted it as true; they have through sheer force of intellect done such violence to their own senses as to prefer what reason told them over that which sensible experience plainly showed them to the contrary."

Not the least of what sensible experience showed men—or perhaps seemed to show them before Galileo instructed them to interpret experience otherwise—was that force is necessary to keep a body in motion. Indeed, where is the experience of inertial motion? It is nowhere. Inertial motion is an idealized conception incapable of being realized in fact. If we start from experience, we are more apt to end with Aristotle's mechanics, a highly sophisticated analysis of experience. In contrast, Galileo started with the analysis of idealized conditions which experience can never know. "Suppose you have a plane surface as smooth as a mirror and made of some hard material like steel, and upon it you have placed a ball which is perfectly spherical and of some hard and heavy material like bronze." Even a perfect sphere on a surface smooth as a mirror was not good enough, and in one of his manuscripts, to make his meaning absolutely clear, he suggested the use of an incorporeal plane. That is to say, Galileo's experiments were mostly conducted on the frictionless planes one meets today in elementary mechanics. They were thought experiments, carried out in his imagination where alone they are possible. Imagine what would be observed, Salviati says to Simplicius, "if not with one's actual eyes, at least with those of the mind." As one modern historian has put it, Galileo got hold of the other end of the stick. Where Aristotle had begun with experience, he began with the idealized case, of which the actual is only an imperfect embodiment. Having defined the ideal, he could then understand the limitations which material conditions, inevitably involving friction, entail. From this point of view, the facts of experience took on new meaning, and many cases, such as projectile motion, which had been anomalies to Aristotle, became comprehensible at once to Galileo. Among the problems solved was the motion of bodies on a moving earth.

At this point in his thought, Galileo made contact with the platonism which animated Copernicus and Kepler. To Galileo, the real world was the ideal world of abstract mathematical relations. The material world was an imperfect realization of the ideal world on which it was patterned. To understand the material world adequately, we must view it in imagination from the vantage point of the ideal. Only in the ideal world do perfectly round balls roll forever on perfectly smooth planes. In the material world, the planes are never perfectly smooth, and the rolling balls, which are never perfectly round, come finally to rest.

Nature is written in code, Galileo said, and the key to the code is mathematics. Kepler could have said as much, and Galileo joined him in accepting an astronomy based on the principle of geometric simplicity. With Galileo, however, the geometrization of nature took a new turn. To Kepler, as indeed to the whole astronomical tradition before him, only the heavenly motions, perfect and eternal, had seemed to offer scope for geometrical analysis. Galileo proposed that geometry be applied to terrestrial motions as well. Such is the ultimate meaning of his assertion that the earth becomes a heavenly body in the Copernican system. If the basic problem to which his work in mechanics addressed itself was posed by the Copernican revolution, the principle of inertia, which he formulated in answer, offered the means to develop a mathematical science of motion such as his youthful work *De motu* had attempted already. The importance he attached to the achievement is reflected in the title he gave to the work that presented it—*Discourses on Two New Sciences** (1638).

One of the two new sciences was dynamics, confined to the single case of the uniformly accelerated motion of heavy bodies falling. Although he refused to discuss what causes heavy bodies to fall and contented himself with describing their motion, he treated free fall in dynamic terms, with a uniform cause producing a uniform effect. When we compare the *Discourses* with *De motu*, it appears that Galileo's advance consisted in seizing on the distinctive feature of a dynamic action. *De motu* had attempted to equate dynamics to fluid statics. The *Discourses* recognized that dynamics must rest on its own principles.

"When I see that a stone, which starting from rest falls from some

---

* In the original Italian: *Discorsi intorno à due nuove scienze.*

height, constantly acquires new increments of velocity, why should I not believe that these additions are made in the simplest and easiest manner of all? The falling body remains the same, and so also the principle of motion. Why should not the other factors remain equally constant? You will say: the velocity then is uniform. [The position of *De motu*.] Not at all! The facts establish that the velocity is not constant, and that the movement is not uniform. It is necessary then to place the identity, or if you prefer the uniformity and simplicity, not in the velocity but in the increments of velocity, that is, in the acceleration."

Clearly the new conception of motion pointed the way to the new understanding of free fall. The approach of *De motu* through fluid statics had expressed the Aristotelian principle that motion itself is an effect requiring a cause. When motion came to be regarded as a state which perseveres unless it is changed, a new effect could be identified. In the passage above, Galileo specified that the dynamic effect of "the principle of motion" (i.e., weight in this case) is acceleration; and because the principle of motion remains constant, the acceleration is constant as well. He concluded further that all bodies, being composed of the same matter more or less densely packed, fall with the same acceleration.

The analysis of fall furnished the prototype of the basic equation of modern dynamics. Galileo himself never considered weight as one example of the broader class we call force, however. To Galileo, weight or heaviness was a unique property of bodies, and he always referred to the tendency of heavy bodies to move toward the center of the earth as their natural motion. He was not alone in his inability to treat gravity as an external force acting on matter, and until scientists learned to do so late in the century, the harvest he had sown could not be fully reaped.

Meanwhile Galileo did succeed in building the foundations of a mathematical science of motion. He defined both uniform motion and uniformly accelerated motion, and described both in mathematical terms. Since geometry represented the very model of science in his opinion, he expressed his results as geometric ratios and not as algebraic equations; but the ratios were equivalent to the basic equations of motion, relating velocity, acceleration, time, and distance, learned today by every beginning student of mechanics.

$$v = at$$
$$s = \tfrac{1}{2} at^2$$
$$v^2 = 2\,as$$

He was able to show as well that bodies experience identical accelerations for all equal vertical displacements. If one body falls freely from rest and another, also starting from rest, descends by an inclined plane through the same vertical distance (which means of course that its path along the inclined plane must be longer and the time for the movement greater), they acquire equal velocities.

The last conclusion played an important role in Galileo's picture of the universe, and it brings us back again to the Copernican system which furnished his cosmology. The circular motion which conserves the integrity of a well-ordered universe is identical to the inertial motion of heavy bodies around a gravitating center. So long as they neither approach the center nor recede from it, no cause operates to change their velocity. Inertial motion, however, can only preserve a velocity; it can never generate one. The motion of heavy bodies toward a gravitating center is the sole natural source of increased velocity, and recession from the center the means whereby motions are destroyed. In both cases, equal increments of velocity correspond to equal radial displacements. For Galileo, the acceleration of gravity was a constant for all distances from the center, just as weight was the constant property of all bodies, however unknown its cause might be.

Between them, Kepler and Galileo confirmed and completed the Copernican revolution. When Galileo died in 1642, probably a minority even among astronomers accepted the heliocentric system. Nevertheless, in the work of Kepler and Galileo its full advantages had been revealed and the major objections to it answered. Its general acceptance had become a matter only of time. The importance of Kepler and Galileo, however, lies less in their relation to Copernicus and the past, than in their relation to the 17th century which followed. In solving the problems of the past, they posed the problems of the future, Kepler in opening the question of celestial dynamics, Galileo that of terrestrial mechanics. In the completion of the work they inaugurated, 17th century science realized its grandest achievements.

CHAPTER II

The Mechanical Philosophy

K EPLER AND GALILEO were not the only scientists of lasting importance at work when the 17th century dawned. In the very year 1600, an English doctor, William Gilbert (1544–1603), published a book entitled *De magnete,** one of the minor classics of the scientific revolution. By universal agreement, Gilbert is recognized as the founder of the modern science of magnetism. His book is revealing in its exposition of the prevailing philosophy of nature.

In its frankly experimental, not to say empirical, approach, *De magnete* stands in marked contrast to the work of Galileo. Galileo regarded experiments primarily as devices by which to convince others; as for himself, he was ready confidently to announce their results without bothering to perform them. Gilbert, on the other hand, undertook to establish the basic facts of magnetism by empirical investigation. From the stories he mentioned, and put to the test, we can learn something of the special awe with which the magnet was regarded; it was the very epitome of the occult and mysterious forces with which the universe was thought to be filled. Stories abounded of such things as magnetic mountains jutting from the sea, which would tear the nails from a ship sailing near. Magnets were said to act as protection against the power of witches. Taken internally (one was allowed first to reduce a loadstone to powder), they were used as a medicine to cure certain diseases. A magnet under the pillow, it was held, drives an adulteress from her bed. (The story was obviously male in origin, and more than good fortune was involved in the apparent immunity of adulterers.) Gilbert took it as his function to winnow fact from fable, and by experi-

---

* *Concerning the Magnet.*

mental investigation to establish the truth of magnetic action. Is it true that diamonds have the power to magnetize iron? Seventy-five diamonds later, Gilbert felt prepared to answer—it is not true.

Gilbert was not the first man to investigate the magnet, and every fact to which he attested was not his own discovery. Nevertheless, the systematic presentation of *De magnete* may be said to have established the basic corpus of facts concerning magnetism. Before Gilbert, magnetic phenomena were frequently confused with static electric phenomena; he distinguished them clearly and definitively. With ample experimental evidence, he demonstrated that the earth itself is a huge magnet, and he insisted that attraction is only one among five magnetic phenomena (or "motions" as he called them). The other four, direction, variation (we say declination), dip and rotation, were all related to the magnetic field of the earth, and assumed greater importance than attraction in Gilbert's eyes.

Gilbert's book, in which so many facts familiar to the student of elementary physics are established on firm evidence, has frequently been hailed as the first example of modern experimental science in action. When we read the work closely, however, and attempt to understand, not solely what modern science has appropriated, but what Gilbert himself maintained, much that is less familiar appears. The title already promises more than the reader from the 20th century expects in a text on magnetism—*Concerning the Magnet, Magnetic Bodies, and the Great Magnet the Earth: a New Physiology Demonstrated both by Many Arguments and by Many Experiments.* A new physiology—that is, a new philosophy of nature—Gilbert saw magnetism, not as one phenomenon among the many which nature displays, but as the key to understanding the whole. The whole, as he understood it, was no less occult and mysterious than the fabled powers of the magnet which he tested so carefully.

Whereas electric attraction is a corporeal action wrought by invisible effluvia, magnetic attraction is an incorporeal power in Gilbert's philosophy. Material bodies do not obstruct it; a magnet attracts iron through glass or wood or paper. If iron can shield a body from attraction, it does so, not by blocking the power, but by diverting it. Especially revealing in his eyes was the ability of a loadstone to excite the magnetic faculty of a piece of iron without suffering any loss in its own potency. Iron (or loadstone, for the two are really identical in his opinion) is genuine telluric matter. Magnetism is its innate virtue, a power it loses only with difficulty and stands ever ready to regain. Utilizing the categories of

Aristotelian metaphysics, he argued that if electricity is the action of matter, magnetism is the action of form. Magnetism is the active principle in primal earth matter.

"Magnetic bodies attract by formal efficiencies or rather by primary native strength. This form is unique and peculiar: it is the form of the prime and principal globes; and it is of the homogeneous and not altered parts thereof, the proper entity and existence which we may call the primary radical, and astral form; not Aristotle's prime form, but that unique form which keeps and orders its own globe. Such form is in each globe—the sun, the moon, the stars—one; in earth 'tis one, and it is that true magnetic potency which we call the primary energy."

As he said in another place, "True earth-matter is endowed with a primordial and an energic form." In perhaps more revealing terms, he identified magnetism as the soul of the earth.

"Attraction" is the wrong word to apply to magnetic action. As Gilbert said, attraction implies force and coercion; it applies properly to electrical action. Magnetic motion, in contrast, expresses voluntary agreement and union. Inevitably the two poles suggested the two sexes, and in language less suited to the Age of the Reformation than to the Restoration, he spoke of the loadstone embracing iron and conceiving magnetism in it. The other magnetic actions seemed more significant to Gilbert than the so-called attraction. Direction, variation, dip—these motions (or rotations) express the underlying intelligence that organizes the cosmos. Gilbert regarded north and south as real directions in the universe, and the magnetic soul of the earth exists to order and to arrange. The compass was "the finger of God," and iron deprived of its magnetism was said to wander lost and directionless. The needle's dip measures latitude; perhaps variation could be used to measure longitude. In Gilbert's fifth motion, revolution, reason itself was ascribed to the magnetic soul of the earth. By "revolution," he referred to the diurnal rotation of the earth upon its axis, a motion he traced to magnetism just as he traced to it the steady direction of the earth's pole as it circles the sun. Placed near the sun, Gilbert asserted, the earth's soul perceives the sun's magnetic field, and reasoning that one side will burn while the other freezes if it does not act, it chooses to revolve upon its axis. It even chooses to incline its axis at an angle in order to cause the variation of seasons.

The first exemplar of modern experimental science turns out to be a very strange book indeed. That is, to the mind of the 20th century it is strange. In the year 1600, however, it must have appeared very familiar because it expressed a prevalent philosophy of nature, what has been called Renaissance Naturalism. To Gilbert, as to many others of his age, nature appeared veritably to pulse with life. The magnetism of primal earth matter corresponded to the active principles present in all things. Matter is never found without life. Neither is it found without perception. As magnetic bodies join in voluntary agreement and union, so sympathies and antipathies, by which likes respond to likes and reject unlikes, relate all bodies one to another. Magnetic attraction indeed was the prime example of the occult virtues that pervaded the animistic universe of Renaissance Naturalism. Gilbert's very empiricism reveals itself as an aspect of the same philosophy. Where Scholastic Aristotelianism had asserted the rational order of nature which the human intellect could probe, the natural philosophy of the 16th century proclaimed the mystery of a nature opaque to reason. Experience, and experience alone, could learn to know the occult forces pervading the universe. As the words "sympathy" and "antipathy" suggest, and as Gilbert's magnetic soul clearly reveals, the occult forces of nature were conceived in psychic terms. Renaissance Naturalism was a projection of the human psyche onto nature, and all of nature was pictured as a vast phantasmagory of psychic forces. Gilbert's *De magnete* was a relatively restrained if unmistakable expression of an established approach to nature.

If the 16th century was the heyday of Renaissance Naturalism, Gilbert was by no means its last representative. Its influence shaped the characteristic conceptions of the Paracelsian chemists of the early 17th century, and in Jean-Baptiste van Helmont (1579–1644) it found a last great figure. It is well known that van Helmont regarded water as the matter from which all things are formed. In a famous experiment, he planted a small tree in a carefully weighed quantity of earth, watered it faithfully, and after it had grown a considerable amount, separated the earth from the roots and weighed it again. The earth had scarcely diminished in quantity, and all of the increased weight of the tree must therefore have derived from the water, converted now to solid wood. In van Helmont's mind, the experiment with the tree fitted neatly into a vitalistic natural philosophy. Water—that is, matter—represents the female principle which requires for its fertilization and animation the male seminal or vital principle. No individual thing is generated in nature, he said, not limiting the statement to what we consider organic today, "but by a getting of the water with childe." Of course, the vital or seminal principle constitutes the ultimate essence of every being, the very source of what it is and does. He referred to it as the image of the master workman, not a dead image but one with "full knowledge" of what it must do and with the power to fulfill itself. The vital principle "doth cloath himself presently with a bodily cloathing;" and molding the matter to the image, it creates the body it animates.

To van Helmont as to Gilbert, magnetic attraction, far from appearing anomalous, represented the very model of action in an animate world. There are, he said, "a Magnetism, and Influential Virtues, every where implanted in, and proper to things." All things are equipped with perception of a sort whereby they perceive those bodies that are like them and those that are foreign—what he called sympathies and antipathies. One of van Helmont's favorite themes was the sympathetic unguent which cures wounds by being applied, not to the wound, but to the weapon which inflicted it. A similar principle explained why the blood of a murdered man runs when the murderer comes near —the spirit in the blood, perceiving the presence of the mortal enemy, boils in rage, and the blood flows. Helmont saw his doctrine as a conscious rejection of materialism, as an assertion of the primacy of spirit. In Aristotelian philosophy, what he referred to in a striking phrase as the "whorish appetite" of matter was given an active role in nature. Quite the contrary, he asserted, the material world "is on all sides governed and restrained by the Immaterial and Invisible."

How can man gain knowledge of the vital principles which constitute the reality of nature? Certainly not by the discursive faculty of reason, which ever falsifies and distorts. "Logick," van Helmont proclaimed, "is unprofitable," and "nineteen Syllogismes do not bring forth knowledge." Instead of reason, which dwells on the surface, understanding alone is adequate to the truth of things. The intellect must be drawn down into the deep; the understanding must transform itself "into the form of the things intelligible; in which point of time indeed, the understanding for a moment is made (as it were) the intelligible thing it self." Things "seem to talk with us without words, and the understanding pierceth them being shut up, no otherwise than as if they were dissected and laid open." Only the understanding, by an immediate intuition of truth, knows things as they are, and knowing things, knows their operations.

In the tradition of Renaissance Naturalism, we are clearly dealing with an ideal of scientific knowledge utterly different from the one we hold. It is the ideal of Faust, the scientist-magician, whose knowledge is of the occult powers of nature.

"Why are we so sore afraid of the name of Magick? [van Helmont asked.] Seeing that the whole action is Magical; neither hath a thing any Power of Acting, which is not produced from the Phantasie of its Form and that indeed Magically. But because this Phantasie is of a limited Identity or Sameliness, in Bodies devoid of choice, therefore the Effect hath ignorantly and indeed rustically stood ascribed, not to the Phantasie of that thing, but to a natural Property; they indeed, through an Ignorance of Causes, substituting the Effect in the room of the Cause: When as after another manner, every Agent acts on its proper Object, to wit, by a fore-feeling of that Object, whereby it disperseth its Activity, not rashly, but on that Object only; to wit, the Phantasie being stirred after a sense of the Object, by dispersing of an ideal Entity, and coupling it with the Ray of the passive Entity. This indeed hath been the magical Action of natural things. Indeed Nature is on every side a Magitianess."

To which Descartes replied in the following terms:

"We naturally have greater admiration for things which are above us than those on the same level or below us. And although the clouds are scarcely higher than the summits of some mountains, nevertheless, because we must turn our eyes toward heaven to look at them, we imagine them to be so elevated that poets and painters see in them the throne of God. All of which leads me to hope that if I explain the nature of clouds in this treatise well enough that there will no longer be any occasion to admire anything that we see in them or that descends from them, it will be readily believed that it is possible in the same way to discover the causes of everything above the earth that appears admirable."

In the 17th century, Descartes spoke for the ascendant school of natural philosophy, whereas van Helmont's voice was one of the last echoes in a fading tradition. Renaissance Naturalism rested ultimately on the conviction that nature is a mystery which in its depth human reason can never plumb. Descartes' call for the abolition of wonder by understanding, on the other hand, voiced the confident conviction that nature contains no unfathomable mysteries, that she is wholly transparent to reason. On this foundation, the 17th century constructed its own conception of nature, the mechanical philosophy.

No one man created the mechanical philosophy. Throughout the scientific circles of western Europe during the first half of the 17th century we can observe what appears to be a spontaneous movement toward a mechanical conception of nature in reaction against Renaissance Naturalism. Suggested in Galileo and Kepler, it assumed full proportions in the writing of such men as Mersenne, Gassendi, and Hobbes, not to mention less well known philosophers. Nevertheless, René Descartes (1596–1650) exerted a greater influence toward a mechanical philosophy of nature than any other man, and for all his excesses, he gave to its statement a degree of philosophic rigor it sorely needed, and obtained nowhere else.

In the famous Cartesian dualism, he provided the reaction against Renaissance Naturalism with its metaphysical justification. All of reality, he argued, is composed of two substances. What we may call spirit is a substance characterized by the act of thinking; the material realm is a substance the essence of which is extension. *Res cogitans* and *res extensa* —Descartes defined them in a way to distinguish and separate them absolutely. To thinking substance one cannot attribute any property characteristic of matter—not extension, not place, not motion. Thinking, which includes the various modes which mental activity assumes, and thinking alone, is its property. From the point of view of natural science, the more important result of the dichotomy lay in the rigid exclusion of any and all psychic characteristics from material nature. Gilbert's magnetic soul of the world could have no place in Descartes' physical world. Neither could the active principles of van Helmont—Descartes' choice of the passive participle, *extensa,* in contrast to the active participle, *cogitans,* which he used to characterize the realm of spirit, served to emphasize that physical nature is inert and devoid of sources of activity of its own. In Renaissance Naturalism, mind and matter, spirit and body were not considered as separate entities; the ultimate reality in every body was its active principle, which partook at least to some extent of the characteristics of mind or spirit. The Aristotelian principle of "form" had played an analogous role in a more subtle philosophy of nature. The effect of Cartesian dualism, in contrast, was to excise every trace of the psychic from material nature with surgical precision, leaving it a lifeless field knowing only the brute blows of inert chunks of matter. It was a conception of nature startling in its bleakness—but admirably contrived for the purposes of modern science. Only a few followed the full rigor of the Cartesian metaphysic, but virtually every scientist of importance in the second half of the century accepted as beyond question the dualism of body and spirit. The physical nature of modern science had been born.

Descartes was fully aware of his revolutionary role in regard to the received philosophic tradition. In his *Discours de la Méthode** (1637), he described his reaction to that tradition as his education had introduced him to it. He had entered upon his education filled with the promise that at its conclusion he would possess knowledge. Far from knowledge, alas, it left him with total doubt. Two thousand years of investigation and argument, he came to realize, had settled nothing. In philosophy, "one cannot imagine anything so strange and unbelievable but that it has been upheld by some philosopher." Descartes decided simply to sweep his mind clear of the past. By a process of systematic doubt, he would subject every idea to a rigorous examination, rejecting everything the least bit dubious until he should come upon a proposition, if such there were, that was impossible to doubt. On such a proposition as a rock of certainty, he could rebuild a structure of knowledge that shared the certainty of its foundation, a structure built anew from the very bottom by reason alone. With the perspective of hindsight, we can see that his repudiation of the past was far less complete than he thought. Nevertheless, his mechanical philosophy of nature was a sharp break with the prevailing conception as represented by Renaissance Naturalism, and scarcely less of a break with Aristotelianism; and in his sensation of making a fresh start he spoke for 17th century science as a whole.

As everyone knows, Descartes found the rock of certainty for which he was searching—that which could not be doubted—in the proposition, *"cogito ergo sum"* (I think, therefore I am). The *cogito* became the foundation of a new edifice of knowledge. From it, he reasoned to the existence of God, and then to the existence of the physical world. In the process of doubt, the existence of a world outside himself had been one of the first items to go; its existence had appeared to depend on the evidence of the senses, and the manifest propensity of the senses to err had called its existence into doubt. From the new foundation of certainty, he now felt able to demonstrate, as a conclusion also beyond doubt, that the physical world external to himself does exist. But to the conclusion he added a condition, perhaps the most important statement made in the 17th century for the work of the scientific revolution. Although the existence of the physical world can be proved by necessary arguments, there is no corresponding necessity that it be in any way similar to the world the senses depict. On the heap of sympathies, antip-

* *Discourse on Method.*

athies, and occult powers already pruned from the physical world were now thrown the real qualities of Aristotelian philosophy. A body appears red, Aristotle had said, because it has redness on its surface; a body appears hot because it contains the quality of heat. Qualities have real existence; they comprise one of the categories of being; by our senses we perceive reality directly. Not so, Descartes retorted. To imagine that redness or heat exist in bodies is to project our sensations onto the physical world, exactly as Renaissance Naturalism projected psychic processes onto the physical world. In fact, bodies comprise only particles of matter in motion, and all their apparent qualities (extension alone excluded) are merely sensations excited by bodies in motion impinging on the nerves. The familiar world of sensory experience turns out to be a mere illusion, like the occult powers of Renaissance Naturalism. The world is a machine, composed of inert bodies, moved by physical necessity, indifferent to the existence of thinking beings. Such was the basic proposition of the mechanical philosophy of nature.

In essays on *La dioptrique* (1637) and *Les météores* (1637), and in the *Principia philosophiae** (1644), Descartes spelled out the details of his mechanical philosophy. One of its foundation stones was the principle of inertia. The mechanical philosophy insisted that all the phenomena of nature are produced by particles of matter in motion —that they must be so produced since physical reality contains only particles of matter in motion. What causes motion? Since matter is by definition inert stuff consciously pruned of active principles, it is obvious that matter cannot be the cause of its own motion. In the 17th century, everyone agreed that the origin of motion lay with God. In the beginning, He created matter and set it in motion. What keeps matter in motion? The very insistence with which the mechanical conception of nature repudiated active principles meant that its viability as a philosophy of nature depended on the principle of inertia. Nothing is required to keep matter in motion; motion is a state, and like every other state in which matter finds itself, it will continue as long as nothing external operates to change it. In impact, motion can be transferred from one body to another, but motion itself remains indestructible.

Descartes attempted to analyze impact in terms of the conservation of the total quantity of motion, a principle which approaches the conservation of momentum formulated later in the century. Since he held that a change in direction alone (without any change in speed)

* *Dioptrics, Meteorology, Principles of Philosophy.*

entails no change in the state of another body, the conclusions at which he arrived vary widely from those we accept. Nevertheless, Descartes' analysis of impact was the starting point of later efforts that bore more fruit. Meanwhile, his rules of impact provided the model of all dynamic action; in a mechanical universe shorn of active principles, bodies could act on one another by impact alone.

It was no accident that the men who constructed the two leading mechanical systems of nature, Descartes and Gassendi, also contributed significantly to the formulation of the concept of inertia. With Galileo, inertia was stated in terms of circular motion corresponding to the diurnal rotation of the earth on its axis. Descartes and Gassendi were the first to insist that inertial motion must be rectilinear motion and that bodies that move in circles or curves must be constrained by some external cause. Such bodies, Descartes asserted, constantly exert a tendency to recede from the center around which they turn. Although he did not attempt to express a quantitative measure of the tendency, his demonstration that such a tendency to recede from the center exists was the first step in the analysis of the mechanical elements of circular motion.

If circular motion ceased to represent perfect motion to Descartes, it continued to play a central role in his philosophy of nature. Although it was not natural, nevertheless it was necessary. Descartes' universe was a plenum. The equation of matter with extension meant that every extended space must, by definition, be filled with matter—or better, must be matter. There can be no vacuum. If there is no empty space into which a body can move, how is it possible that there be any motion at all? It is possible, Descartes replied, only because every body that moves moves into the space that it vacates, as it were, at the same time. Put in other terms, every moving particle in a plenum must participate in a closed circuit of moving matter, like the rim of a wheel turning on its axis. Hence every motion must be circular—although, of course, the word "circular" in this context refers to a closed orbit of some shape, not to the perfect circle of Euclidean geometry. Because circular motion, though necessary, is unnatural, it sets up centrifugal pressures in the plenum. Descartes traced the major phenomena of nature to such pressures.

The first consequence of the introduction of motion into the infinite plenum that is our universe is the establishment of an infinite number of vortices. Descartes pictured the vortex in which our solar system is located as a whirlpool of matter so huge that the orbit of Saturn is to the whole no more than a point. Most of the vortex is filled with tiny balls turned into perfect spheres by the incessant bumping of one on another. These he referred to as the "second element." The "first element," the "aether" as it was often referred to in the 17th century, is composed of the extremely fine particles which fill up the spaces between the spheres of the second element and all other pores as well. There is also a third form of matter in Descartes' universe, bigger particles which are collected into the large bodies we call planets. As the whole vortex whirls about its axis, every particle in it endeavors to recede from the center, but in a plenum one particle can move away from the center only if another moves toward it. Like every other body, each planet tends to recede from the center, but at some distance from the center its tendency to recede is just balanced by the tendency of the swiftly moving matter of the vortex beyond it. An orbit is established by the dynamic balance between the centrifugal tendency of a planet and the counterpressure arising from the centrifugal tendency of the other matter composing the vortex.

The vortical theory constituted the first apparently plausible system designed to replace the crystalline spheres. To be sure, Kepler's celestial mechanics had preceded it, but Kepler's system had been constructed on principles unacceptable to the mechanical philosophy. Descartes' vortex, needless to say, was acceptable, and for half a century it dominated physical accounts of the heavens. To understand scientific thought in the 17th century, it is important to realize what it pretended to explain and what it did not pretend to explain. The vortex offered a mechanical account of the gross celestial phenomena. It explained why the planets are carried about the sun, all in the same direction and all in (about) the same plane. By the covert introduction of arbitrary factors, it explained why the planets move more slowly the further they are removed from the sun. These things it explained, moreover, as the necessary consequences of matter in motion, without recourse to any occult powers. To science in the 17th century, the type of mechanical explanation that the vortex offered was important, and it is not difficult to understand the theory's appeal. What the vortex made no attempt to treat were the precise details of planetary orbits which constituted the domain of technical astronomy. Kepler's three laws were not mentioned by Descartes, and it is hard to see how he could have derived them from the vortex. But the sort of mathematical description that Kepler's laws represent was also important to 17th century science. The mechanical philosophy, with its concentration on physical causation, existed in tension with the Pythagorean tradition of mathematical description. The highest achievement of science in the 17th century, the work of Isaac Newton, consisted in the resolution of that tension.

The solar system was not the sole topic of Descartes' philosophy of nature. It was also not the most difficult. As its fundamental proposition, the mechanical philosophy asserted that all the phenomena of nature are produced by inert matter in motion. What about light? No philosophy of nature that ignores light can pretend to be complete, and light appears to be the least obviously mechanical of all phenomena. In Descartes' system, however, it stands revealed as a necessary mechanical consequence of the vortex. The sun is the principal source of light in our system, and the sun is also at the center of the vortex. We have already seen that circular motion sets up centrifugal pressures throughout the vortex, and the physical reality of light is nothing more than such pressure. Received on the retina of the eye, it causes a motion in the optic nerve which in turn produces the sensation we call "light." Moreover, Descartes added, since pressure is a tendency to motion, it obeys the laws of motion, and the laws of reflection and refraction can be shown to follow as necessary consequences.

Gravity (i.e., *gravitas,* the heaviness of bodies near the surface of the earth) scarcely appears more mechanical in origin than light. To explain it, Descartes posited a small vortex around the earth, turning with the earth and terminating at the height of the moon. Again the centrifugal tendencies inherent in circular motion were called upon, and again the necessities of the plenum. What is gravity? It is a deficiency of centrifugal tendency by which some bodies are forced down toward the center by others, with a greater centrifugal tendency, which rise. It emerged as a regrettable consequence of Descartes' theory that bodies should fall, not along the perpendicular to the surface of the earth, but along the perpendicular to the axis. Mechanical philosophers, who were concerned to reveal the cause of every phenomenon, had to learn to tolerate minor discrepancies.

Perhaps the crucial case for the mechanical philosophy of nature was magnetism. To an earlier age, it had represented the very epitome of an occult power. Correspondingly, the mechanical philosophy had to explain away magnetic attraction by inventing some mechanism that would account for it without recourse to the occult. Descartes' was particularly ingenious. In considerable detail, he described how the turning of the vortex generates screw-shaped particles which fit similarly shaped pores in iron. (See Fig. 2.1.) Magnetic attraction is caused by

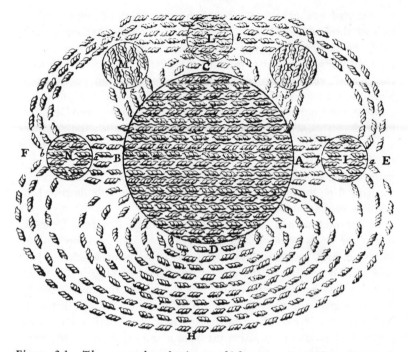

Figure 2.1. The screw-shaped pieces which cause magnetic action pass through the earth and through five loadstones shown in various positions as they align with the earth's magnetic field.

the motion of the particles, which in passing through the pores in magnets and iron, drive the air from between the two and cause them to move together. What about the fact of two magnetic poles? Very simple, Descartes replied; there are left-handed screws and there are right-handed screws.

The treatment of magnetism is revealing of the basic motivations of Cartesian science. In contrast to Gilbert, Descartes did not undertake a detailed investigation of magnetic phenomena. He regarded the phenomena as given; there was no need to confuse himself by searching for more. The problem was not the phenomena but their interpretation, and Descartes' purpose was to demonstrate that there are no magnetic phenomena which cannot be explained in mechanical terms. In the same way, when his *Principles of Philosophy* came to the detailed discussion of nature, Descartes assumed that the phenomena were known. His science was not devoted to careful investigations of nature, not to the discovery of new phenomena, but to the elaboration of a new explanation of those already known. There is no necessity that the physical world be similar in any way to the one our senses depict; it consists solely of particles of matter in motion. Descartes' purpose was to show that for all known phenomena causal mechanisms can be imagined. Since the mechanical philosophy as such offered no criteria of what is possible, some rather strange phenomena found their way into Descartes' universe. Helmont's discussion of blood running when the murderer approaches strikes us as the epitome of absurdity; Descartes accepted the fact and imagined an effluvial mechanism to explain it. The sympathetic unguent did not appear in his work, but Kenelm Digby, a mechanical philosopher of the following generation, duly described the invisible mechanism by which it cures.

Earlier philosophies had seen nature in organic terms. Descartes turned the tables by picturing even organic phenomena as mechanisms. In his universe, man was unique—the one living being which was both soul and body. Even in the case of man, however, the soul was not considered to be the seat of life, and all organic functions were described in purely mechanistic terms. The heart became a tea kettle, its heat analogous to the heat of fermentation (in itself a mechanical process to Descartes), its action the boiling and expansion of the drops of blood which were forced into it from the veins and forced on by the pressure of vaporization. Other animals, lacking a rational soul, were nothing but complicated machines. If there were automata, Descartes asserted, "possessing the organs and outward form of a monkey or some other animal without reason, we should not have had any means of ascertaining that they were not of the same nature as those animals."

Many of Descartes' explanations of phenomena differ so widely from those we now believe to be correct that we are frequently tempted to scoff. We must attempt rather to understand what he was trying to do and how it fit into the work of the scientific revolution. The cornerstone of the entire edifice of his philosophy of nature was the assertion that physical reality is not in any way similar to the appearances of sensation. As Copernicus had rejected the commonsense view of an immovable earth, and Galileo the commonsense view of motion, so Descartes now generalized the reinterpretation of daily experience. He did not intend to conduct the sort of scientific investigation we are familiar with today. Rather his purpose was metaphysical—he proposed a new picture of the reality behind experience. However wild and incredible we find his explanations, we must remember that the whole course of modern science has been run, not by returning to the earlier philosophy of nature, but by following the path he chose.

Certainly the 17th century found the appeal of the mechanical philosophy of nature overwhelming. The mechanical philosophy did not mean solely the Cartesian philosophy, however, and among other mechanical approaches to nature, one at least stood as a viable and attractive alternative, Gassendi's atomism. Inevitably, the atomic philosophy of antiquity had reappeared in western Europe with the general recovery of ancient thought during the Renaissance. Galileo had felt its influence, and its mechanistic treatment of nature probably helped to shape Descartes' system. It remained, however, for a contemporary of Descartes, Pierre Gassendi (1592–1655), to espouse and expound atomism as an alternative mechanical philosophy. As a thinker, Gassendi was utterly unlike Descartes. Where Descartes saw himself as a systematic philosopher rebuilding the philosophic tradition on new principles of his own creation, Gassendi considered himself as a scholar drawing together the best elements that the tradition could offer. His principal work, *Syntagma Philosophicum** (1658), is an unreadable compilation of everything ever said on the topics discussed, a compilation further which intended to exhaust discussable topics. The work grew like Topsy, and was published in its ultimate form only as a posthumous work, when the author was finally beyond the possibility of adding and patching. In a word, Gassendi was the original scissors and paste man, and his book contains all the inconsistencies of eclectic compilations. At least three different conceptions of motion are put forward in it with no effort whatever to reconcile them. From the tradition one system appealed to him above the others, however, and the *Syntagma* was unmistakably an exposition of atomism.

Being an atomist, Gassendi differed from Descartes on certain specific questions. Descartes argued that matter is infinitely divisible; Gassendi of course maintained that there are ultimate units which are never divided. The very word "atom" derives from the Greek word for indivisible. Descartes' universe was a plenum; Gassendi in contrast argued for the existence of voids, spaces empty of all matter. Both issues are important philosophic questions, but the disagreements of the two men pale beside their large areas of agreement. They asserted

---

* *Philosophical Treatise.*

alike that physical nature is composed of qualitatively neutral matter, and that all the phenomena of nature are produced by particles of matter in motion.

Far more important for later science was another difference between Descartes and Gassendi which was logically connected with the question of the plenum. Descartes' insistence that nature is a plenum was the necessary consequence of his identification of matter with extension, and the identification of matter with extension in turn made possible the utilization of geometric reasoning in science. Because geometric space is equivalent to matter, natural science might hope to attain the same rigor in its demonstrations that geometry was agreed to have. Indeed his method, four rules to govern investigations, was little more than a restatement of the principles of geometric demonstration. Rebel against the prevailing tradition though he was, Descartes accepted an ideal of science that went back to Aristotle. It held that the name "science" applies, not to conjectures, not to probable explanations, but solely to necessary demonstrations rigorously deduced from necessary principles. If such a degree of certainty could not be attained in the details of causal explanations, where it was possible to imagine more than one satisfactory mechanism, at least the general principles were beyond doubt—the rigorous separation of the corporeal from the spiritual, and the consequent necessity of mechanical causation.

When Gassendi denied the equation of matter with extension, he denied as well the program of Cartesian science. Atoms are extended, but extension is not their essence. He was convinced indeed that knowledge of the essence of things is beyond the reach of finite man. Gassendi accepted a degree of skepticism as an inevitable ingredient of the human condition. God and God alone can know ultimate essences. Hence the ideal of science held by the dominant school of philosophy in the western tradition from Aristotle to the 17th century and reaffirmed by Descartes was labelled an illusion. Thoroughgoing skepticism was not Gassendi's conclusion, however; he offered instead a redefinition of science. Nature is not completely transparent to human reason; man can know her only externally, only as phenomena. It follows that the only science possible to man is the description of phenomena, a new ideal of science which found its earliest statement in Gassendi's logical writings. Implicit already in Galileo's description of the uniform acceleration of free fall whatever its cause, the ideal was stated formally by Gassendi as part of his denial of the traditional one. It was not an easy conception to grasp, and mechanical philosophers in the 17th century continued to imagine microscopic mechanisms to "cause" natural phenomena. In Isaac Newton, however, Gassendi found a follower, and in the work of Newton, his definition of science demonstrated what it could foster. It has become so deeply ingrained in the procedures of modern experimental science that we find it difficult today to comprehend the Cartesian (and Aristotelian) ideal of necessary demonstrations —although that ideal appeared self-evident to men before the 17th century.

Gassendi's discussions of method were one thing; Gassendi's practice was something else. In the bulk of his work, where he took up the details of natural philosophy, fine phrases about restricting science to the description of phenomena could not restrain him from the occupational vice of mechanical philosophers, the imaginary construction of invisible mechanisms to account for phenomena. In many ways, the qualitative philosophy of Aristotle reappeared in disguise in his writings; that is, special particles with special shapes were posited to account for specific qualities. Descartes equated heat with the motion of the parts of bodies and considered coldness as the absence of heat. Gassendi, on the other hand, spoke of calorific and frigorific particles. Nevertheless, by insisting on particles and allowing differences solely in shape and motion, he maintained allegiance to the basic principles of the mechanical philosophy of nature. Robert Boyle, a leading mechanical philosopher as well as chemist of the following generation, treated atomism and Cartesianism as two expressions of the same conception of nature. We owe the name, "mechanical philosophy," to Boyle. As he summed it up, the mechanical philosophy traces all natural phenomena to the "two catholic principles," matter and motion. He might have added that by "matter" the mechanical philosophy means qualitatively neutral stuff, shorn of every active principle and of every vestige of perception. Whatever the crudities of the 17th century's conception of nature, the rigid exclusion of the psychic from physical nature has remained as its permanent legacy.

Meanwhile, in the 17th century, the mechanical philosophy defined the framework in which nearly all creative scientific work was conducted. In its language questions were formulated; in its language answers were given. Since the mechanisms of 17th century thought were relatively crude, areas of science to which they were inappropriate were probably frustrated more than encouraged by its influence. The search for ultimate mechanisms, or perhaps the presumption to imagine them, diverted attention continually from potentially fruitful enquiries and hampered the acceptance of more than one discovery. Above all, the demand for mechanical explanations stood in the way of the other fundamental current of 17th century science, the Pythagorean conviction that nature can be described in exact mathematical terms. Despite its rejection of a qualitative philosophy of nature, the mechanical philosophy in its original form was an obstacle to the full mathematization of nature, and the incompatibility of the two themes of 17th century science was not resolved before the work of Isaac Newton. Meanwhile, virtually no scientific work in the 17th century stood clear of its influence, and most of the work cannot be understood apart from it.

CHAPTER III

Mechanical Science

THE PROMINENCE which a set of long-known phenomena suddenly acquired in the middle of the 17th century can be attributed to the rise of the mechanical philosophy and mechanical modes of explanation. The cupping glass, a glass heated and placed over a sore, was an ancient instrument for drawing infected matter. It was known likewise that water does not run out of a narrow-necked bottle when it is filled and inverted. The operation of pumps and syphons was analogous. In their case, perhaps, an effect appeared that was disturbingly non-analogous. Pumps would not draw water more than about thirty-four feet and syphons would not operate over hills of more than that height. In both cases, however, it was universally agreed that imperfections in the materials caused the failure. Since the pipes in use were of wood, the conclusion was not without apparent justification. In the established philosophy of nature, all of the phenomena were referred to nature's abhorrence of a vacuum, an explanation which embodied the principles the mechanical philosophy had been created to destroy. It implied that nature has sensitive and active faculties by which she perceives threats to her continuity and moves to oppose them. For such phenomena, moreover, alternative mechanical explanations were obvious.

A passage in Galileo's *Discourses,* published in 1638, effectively started the debate. As part of his analysis of the breaking strength of beams, Galileo needed a theory of the cohesion of bodies. The observed fact that a syphon carries water over a maximum height of about thirty-four feet seemed to offer a foundation on which to build. Above all, it provided an exact quantitative factor, the weight of a unit column of water some thirty-four feet high. He attributed the column of water to what he called the attraction of the vacuum; and arguing that bodies are composed of infinitesimal particles separated by infinitesimal vacua, he went on to construct a theory of cohesion from the attraction of the vacuum. Galileo's explanation of cohesion never enjoyed success, but the publication of the *Discourses* injected the phenomena on which it was based into the current of scientific discussion.

Among others who considered it were a scientific circle in Rome. Here the proposal was made in effect to isolate one leg of a syphon, and thus the first barometer was constructed in the early 1640s—a water barometer with a glass bulb at the top. The water stood at about thirty-four feet, its upper surface visible through the glass. What was above the water? It was not apparent that anything was above the water. At least some men who observed it argued that the space was a vacuum, and maintained that the weight of the atmosphere sustained a column of water equal in weight. An obvious means to check the explanation suggested itself. Sea water is heavier than fresh water; if sea water were substituted, the height of the column ought to be less. It was Torricelli (1608–1647), a young admirer of Galileo, who suggested the use of a different fluid which is immensely heavier than water; and in 1644, he constructed the first mercury barometer. The name "barometer," which suggests an instrument to measure the pressure of the atmosphere, is of course a misnomer applied to Torricelli's tube. Instead of measuring the pressure or weight of the atmosphere, he was using the atmosphere taken as constant to measure the weight of the fluid column enclosed in the tube. The first barometer was an experiment as nicely calculated as any experiment in the history of science. If a simple mechanical balance is in play, with the atmosphere on one side and the enclosed fluid on the other, the substitution of mercury, which is fourteen times as dense as water, should yield a column one fourteenth as high. When the column of mercury in Torricelli's tube stood at twenty-nine inches, he had confirmed the mechanical explanation. Although at least two decades of argument and experimentation were necessary before the mechanical explanation was generally accepted, Torricelli's construction of the first mercury barometer appears now to have confirmed it beyond any reasonable doubt.

Discussions of the barometer inevitably involved the vacuum. Arguments against the existence of a vacuum were well-established, deriving as they did from the text of Aristotle. On the one hand, the argument from motion said that in a vacuum resistance would be zero and velocity therefore infinite; on the other hand, the argument from logic asserted that the existence of a vacuum, that which is "no thing" (the same pun existed in Greek), would be a contradiction in terms. The barometer itself had not been known to Aristotle, of course, and in discussing it, the Aristotelians were thrown back on their own resources. Since something had to occupy the space, one school suggested that a bubble of air must be present; when the tube is set up, the bubble expands, or better is stretched, until its tension is sufficient to sustain the mercury. Another school of thought held that a vapor forms above the liquid, driving it down: without the vapor, the mercury would fill the tube entirely. The explanations were obviously *ad hoc*. The very fact suggests the golden opportunity which the barometer offered to the mechanical philosophy. A simple phenomenon with a quantitative factor, it presented the most favorable ground on which the mechanical philosophy might attack animistic conceptions. The quantitative factor, moreover, made the question ideally suited for experimental investigation. Because of the quantitative factor, it was possible to design experiments to test the *ad hoc* Aristotelian explanations one after another, and when the argument over the barometer was settled, it had furnished a classic example of the power of experimental investigation.

Blaise Pascal (1623–1662) played the most important role in the experimental demonstration. A young man just reaching maturity, he had the good fortune to take up the question at the time and in the place where technical progress in glass blowing made it possible to carry out the experiments he devised. Rouen, where he lived, was a leading center in the manufacture of glass, and for the first time pipes as long as fifty feet were available, enabling Pascal to carry out experiments either with mercury or with water. The argument that vapors drive the column down, whereas it would otherwise stand at the top of the tube, suggested a comparison of water with wine. By universal agreement, wine was held to be the more spirituous liquid and therefore more productive of vapors. On the other hand, wine is lighter than water; and if the mechanical explanation were correct, the column of wine should stand higher. On a famous occasion in the harbor of Rouen, Pascal had two long pipes erected beside the mast of a ship, one filled with wine and one with water. The audience had been asked to predict the result before the experiment was performed, and the supporters of the spirituous theory saw it destroyed before their eyes.

In a similar way, Pascal devised experiments which submitted the other peripatetic explanations to the trial of quantitative examination. If it were true that a bubble of air were included in the tube and sustained the column of liquid by its tension, then a relation between the length of the column and the space above it should be found. Pascal erected a mercury barometer in a tube fifteen feet long and in another with a huge bulb at the end. (See Fig. 3.1.) In both cases, and in every other case he tried, the constant factor was the length of the column of mercury whatever the size of the space above. If he inclined the tube, moreover, the vertical height of the surface remained constant, so that the space above the mercury could be made to shrink away to nothing, leaving no visible bubble at all when the end of the tube descended below twenty-nine inches.

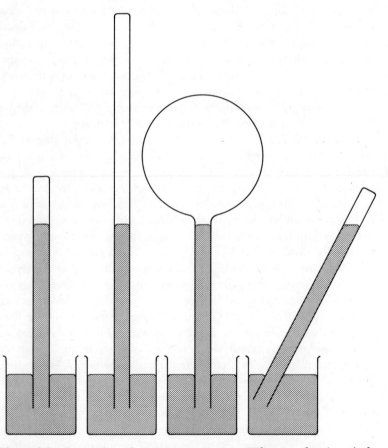

Figure 3.1. Proportion of vacuum to mercury. Whatever the size of the space above the mercury, the height of the column continues to stand at about twenty-nine inches.

In his early work on the Torricellian vacuum, Pascal drew from the experiments a conclusion which appears surprising, not to say dumbfounding, to the 20th century reader. The conclusion was that nature abhors a vacuum. A second conclusion substantially modified the first, however,—to wit, that nature's abhorrence of a vacuum is finite and measured by the weight of a unit column of mercury twenty-nine inches high. When a greater force is applied, a vacuum (or at least a space devoid of tangible matter) can be created. What appears as a compromise in fact demanded more than the Aristotelian philosophy could concede because it admitted the possibility of a vacuum under certain conditions. Pascal's true opinion went well beyond, but he was concerned with the degree of validity of inductive conclusions. Experimenting in an age before the invention of the air pump, he was unable to vary the weight on one side of the balance he believed to be in play, and hence the weight on the other side remained constant. Until he could vary the weight, he felt that the evidence validly supported only the conclusion that nature's abhorrence of a vacuum is finite and measured by the weight of the unit column.

In the end, Pascal thought of a way to vary the weight and to demonstrate the conclusion he fully believed. If he could not alter the atmosphere, he could alter the depth of the barometer in it. His brother-in-law lived in central France near the mountain called the Puy de Dôme. Pascal asked him to perform the experiment. While one barometer was left at the foot of the mountain as a control, another was carried to the summit. Of course, the height of the barometer at the summit dropped.

The experiment on the Puy de Dôme is one of the most famous in the entire history of science. By careful definition of the conditions, Pascal contrived an experiment to bring the issue under discussion directly to test, and the outcome sustained his conclusion that in the barometer a simple balance of weights is in play, the weight of the atmosphere and the weight of the column of liquid. Less well known but no less brilliant in its design was the experiment Pascal called the vacuum in a vacuum. A tube was blown and bent so as to leave two vertical legs, each something over twenty-nine inches long, end to end, with a large bulb between the two where a reservoir of mercury could collect. (See Fig. 3.2.) When the entire apparatus was filled and erected in a dish of mercury, the lower leg functioned as an ordinary barometer. No atmosphere bore down on the surface of mercury in the central reservoir, however, and the mercury in the upper tube stood no

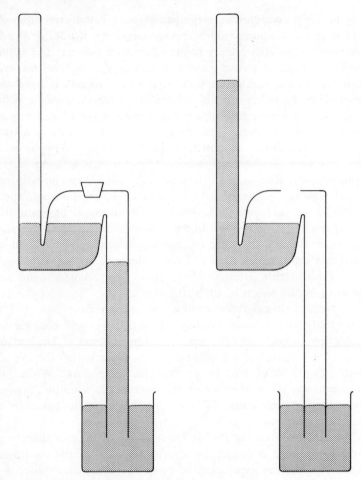

Figure 3.2. Pascal's vacuum in a vacuum.

higher than the surface of the reservoir. A hole at the top of the lower leg, closed with a plug, could be opened to admit air slowly. As the air entered, the mercury in the lower leg dropped while that in the upper leg rose until, when the hole was fully opened to the atmosphere, the upper leg functioned as an ordinary barometer, and the mercury in the lower leg stood at the same level as that in the dish.

After Pascal's experiments, it was no longer possible to contend intelligently that the barometer does not function as a simple mechanical balance. In his treatment, the weight of the atmosphere alone was the determining factor of the height of a given fluid. The invention of the air pump in the 1650's led to the development of a further conception by Robert Boyle (1627–1691). When a barometer was enclosed in the receiver of the pump, the height of its column of mercury initially stood at the normal height, but fell as the air was pumped out. There could be no question of a balance of weights, for the weight of the air within the receiver was not a fraction of the mercury's weight. In another experiment, a bladder containing a small amount of air expanded continually as the receiver was exhausted. Such phenomena led Boyle to suggest that air is an elastic fluid which expands when external constraint is removed. Because of its elasticity air exerts pressure, and the pressure of the air rather than merely its weight sustains the barometer's column of mercury. In the open air, of course, the weight of the atmosphere maintains the pressure, but when a barometer is enclosed in a bell jar, the column continues to stand at twenty-nine inches because the receiver, which bears the weight of the atmosphere, also maintains the pressure in the confined air. Boyle referred to elasticity as the "spring in the air," and as a good mechanical philosopher he imagined that each particle of air is a little spring which can be compressed by an external force.

The publication of Boyle's experiments and conjectures in *New Experiments Physico-Mechanical, Touching the Spring of the Air* (1660) provoked a counterattack by an English Jesuit, Father Linus. Rejecting the concept of elasticity, Linus pointed out the apparently absurd consequence of Boyle's concept, namely that air should be subject to further compression as well as to expansion. The challenge was Boyle's invitation to immortality, for the resulting investigation ended in the statement of Boyle's Law. By bending the closed end of a long glass tube into a U shape and confining some air in it above mercury, he could subject the air to pressures of several atmospheres by pouring mercury into the other leg. The volume of the air was easily measured by the length of the space it occupied, and the reciprocal relation of pressure and volume, suspected before the experimentation began, emerged at once.

Boyle's law is an ideal product of 17th century science. A simple quantitative relation, it satisfied at once the search for exact mathematical descriptions of phenomena and the demand for mechanical explanations. The mechanical philosophy could not have hoped to find more advantageous ground than fluid statics offered on which to attack the prevailing philosophy of nature. The basic relations of the lever and balance had been known since antiquity. The analogy of the barometer was seen at once, and any number of experiments repeating the simple quantitative relations could be devised. In contrast, the nonmechanical explanations with their implicit animism could never account adequately for the quantitative aspect of the phenomena. They were clearly *ad hoc* explanations, and the superiority of the mechanical one was beyond serious question.

Optics is less obviously suited than fluid statics to mechanical modes of thought. Nevertheless the study of optics, a science to which the 17th century devoted a great deal of attention, was profoundly influenced by the mechanical philosophy. Inevitably the nature of the influence differed. In the case of the barometer, it fostered the recognition of the essential factors in a purely mechanical equilibrium. In the case of optics, the mechanical philosophy encouraged the generation of mechanical conceptions of light which might account for the known phenomena. It is difficult to argue that the mechanical philosophy led to any discoveries in optics, and it may have frustrated the comprehension of some. It is beyond question, however, that it provided the idiom in which the 17th century discussed optics.

As far as the first great figure in 17th century optics is concerned, the assertion goes beyond the truth. Johannes Kepler took up the study of optics as an aspect of astronomy, entitling his great work *Astronomiae pars optica** (1604). In it, he established the fundamental propositions on which the study of optics has rested ever since. Kepler was concerned primarily with the physiological problem of sight. Ancient optics had approached the question in terms of a visual pyramid, the base of which rested on the object perceived, with the apex in the eye. (See Fig. 3.3.) In the atomist philosophy, which rendered the concept in material terms, objects continually give off simulacra of themselves, thin films of atoms which reproduce the object in its shape and color. Shrinking along the lines of the visual pyramid, the simulacra enter the eye where they are perceived. Whether light was considered to be something which emanates from the eye or something external which enters it, all schools had agreed that objects are seen in their organic unity in the act of vision.

The essence of Kepler's reform in optics, a reform in which he drew on the Arab, Alhazen, and the medieval student of optics, Witelo, lay in breaking down the object of vision into an infinite number of points.

---

* *The Optical Part of Astronomy.*

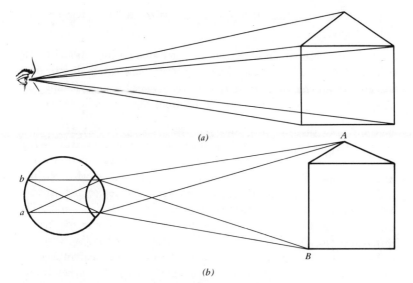

Figure 3.3. (a) *The optical pyramid.* (b) *Kepler's theory of vision.*

Light, Kepler said, has the property of flowing out from point sources in an infinite number of straight lines which we call rays. Every point of a visible object can be thought of as a point source, and the basic problem of optics is to trace a pencil of rays diverging from some point until they are brought to a focus at another. Kepler's optics was summed up in ray-tracing, and ray-tracing led him to reverse the optical pyramid. (See Fig. 3.3.) From any point of a visible object as an apex, a particular pyramid (or, better, cone) of rays has the pupil of an eye as its base. Inside the eye, a second cone on the same base has its apex on the retina. There a point image of the point in the visible object is formed, and a pattern of point images comprises our vision of bodies.

By an extension of the same analysis, Kepler was able to solve the basic problems presented by reflection and refraction. Why does an object seen in a mirror appear to be placed behind it? The eye that receives the pencils of rays from the points of the object lacks any means of perceiving the path which the rays have followed before they reach it. As far as the eye is concerned, rays only travel on straight lines. It focuses the pencils and places the object as though the entire paths of the rays had followed the lines of the final segments. Hence the eye places the object behind the mirror. A similar analysis explains apparent displacement by refraction. In the late 16th century, the

Italian, della Porta, had attempted to discuss refraction in terms of the old concepts.

"If an eye observes an object under water along a line normal to the surface, the object [sic] leaps from the water and enters directly into the eye; if on the contrary it is observed along an oblique line, the object leaps from the water but inclines from the perpendicular."

In contrast to the utter inadequacy of such concepts for a satisfactory analysis, Kepler's appear to provide the basis necessary for clarity. When he wrote the Astronomiae pars optica, the telescope was unknown, and Kepler did not discuss lenses as such. Seven years later, when Galileo had made the telescope an object of attention, Kepler published a second work, Dioptrice* (1611), devoted to the theory of lenses. The theory remained incomplete because he failed to discover the law of refraction. (Refraction is the change of direction or bending that rays of light undergo when they pass obliquely from one transparent medium into another—in this case, from air into glass.) Nevertheless, his Dioptrice became the foundation of all later work on lenses, as his first book was for the science of optics as a whole.

In breaking down the unity of visible objects into points, which assume greater reality as far as optics is concerned than the object considered as a whole, Kepler employed one of the basic concepts of the mechanical philosophy's approach to nature. Points stand in the same relation to visible objects as particles or atoms in the mechanical philosophy to ordinary bodies in the world of experience. With Descartes, more than analogous modes of thought entered optics. It was essential to the completeness of his philosophy of nature that he bring light under its general principles. Indeed, he did more than bring it under his general principles. He treated light from the sun as a necessary consequence of matter in motion in a vortex.

When he came to write his own Dioptrics (1637), Descartes had to make his general remarks on light a great deal more specific. Essentially, his conception treated light as a pressure transmitted instantaneously through transparent media. In the Dioptrics, he used the analogy of the stick with which a blind man "sees." When one end of the stick hits a stone, the motion of the end is transmitted through the stick to the hand, and the blind man "sees" the obstacle in his way. Since nature is a plenum, we may think of transparent media as solid

* Dioptrics.

matter resting against the eye. A pressure arising in a luminous body makes an impression on the retina, causing a motion in the optic nerve which is transmitted to the brain and interpreted as light. Fully to explain light, Descartes also used two other mechanical analogies, the second of which compared light to the motion of a tennis ball. Because pressure is a tendency to motion, he declared, it obeys the same laws as motion. Without further ado, he set about using this analogy to derive the laws of reflection and refraction.

The law of reflection followed easily enough from the example of the tennis ball. The rectilinear propagation of light corresponds to the inertial motion of the ball after the racket hits it, and by analyzing the motion into a component parallel to the reflecting surface, which is not altered by the bounce, and one perpendicular which is reversed, he demonstrated easily that the angle of reflection is equal to the angle of incidence. (See Fig. 3.4.) Since the law of reflection had been known for centuries, the demonstration was hardly a triumph, whatever we may think of its rigor.

Refraction was another matter, however; its law, if one there were, remained unknown. Descartes approached refraction with the same principles, replacing the reflecting surface with a cloth, which represents the refracting interface of two media, through which the ball passes. (See Fig. 3.4.) Suppose the second medium is one in which light travels more slowly than in the first. All of the change in velocity takes place at the surface, Descartes argued, and all of the change is associated with the vertical component by which the ball breaks through. By drawing out the consequences of the premises, he proceeded to show that for all angles of incidence at which light is refracted into the second medium, the sine of the angle of incidence is proportional to the sine of the angle of refraction.

As a demonstration, the argument was preposterous. Its foundation consisted of assumptions which were arbitrary and contradictory. He held that pressure, which he called a tendency to motion, is subject to the same laws as motion itself. In the case in which light travels more swiftly through the second medium, he had to imagine that the tennis ball receives a second stroke as it passes through the cloth, a device for which the optical counterpart is difficult to imagine. Above all, the demonstration required that light travel at different velocities in the two media, whereas Descartes maintained elsewhere that the movement of light is instantaneous. Although experimental testing of the law would not have been difficult, Descartes did not undertake to carry it out. The

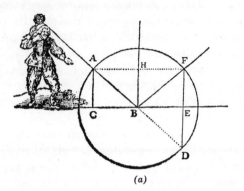

(a)

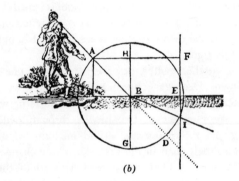

(b)

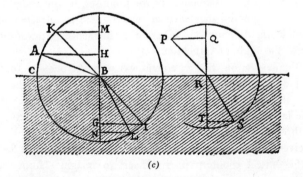

(c)

Figure 3.4. (a) *Reflection.* (b) *Refraction.* (c) *The law of refraction. For a given pair of transparent media, KM/LN = AH/IG. That is, for any angle of incidence, (sin i)/(sin r) = n, a constant for that pair of media.*

demonstration was preposterous—but the result is still accepted as the sine law of refraction. Fermat later placed the sine law on a different footing by showing that light follows the quickest path between two points in different media when it is refracted according to the sine law.

One school of thought explains the anomaly of Descartes' demonstration by plagiarism. We know that a Dutch scientist, Snel, also discovered the law of sines, and since Descartes was resident in the Netherlands, he could have seen Snel's unpublished papers. There is no evidence to support the charge, however, and it appears far more likely that mathematical research led him to the discovery. Use of the telescope had made it plain that spherical lenses do not refract parallel rays to a focus, and Descartes was interested in discovering the "anaclastic" curve, the shape of the surface that refracts to a focus. What more natural than to try a conic section since a parabolic mirror was known to reflect to a focus? By investigating the ellipse and the hyperbola, Descartes may have discovered what he demonstrated in the *Dioptrics,* that if light is refracted according to the sine law, elliptical or hyperbolic lenses will focus parallel rays. Since no one could grind a true elliptical or hyperbolic lens to test it, the demonstration in itself did not constitute a serious reason for accepting the sine law. Another demonstrated consequence did. In his treatise *Les météores,** he showed that the primary rainbow could never be seen higher than $41°47'$ or the secondary rainbow below $51°37'$. The demonstration rested on the sine law, and observations confirmed it.

Descartes also attached the phenomena of colors to the science of optics. Hitherto, light and color had been considered as two different things. Colors were real qualities of bodies, illuminated by light but distinct from light. Not all colors, however, for there were phenomena, such as the rainbow, in which colors that were obviously not on the surfaces of bodies appeared. These were called "apparent colors" to distinguish them from real colors, and they were assigned to the modification which light undergoes in passing through a dark medium. Descartes' philosophy denied the possibility of real qualities such as colors. By definition, all colors are only appearances, and it was incumbent on him to show that the appearances can be traced to the same principles assigned to light. Light is a pressure transmitted through a medium composed, he held, of tiny spheres. Colors, obviously, are the

---

* *Meteorology* is the most satisfactory, if not entirely accurate, translation.

sensations caused by a tendency to the other motion such spheres can have, namely rotation on their axes. By an involved argument based on an experiment with a prism, he concluded that refraction can alter the rate of rotation; an increased rotation causes the sensation of red, and a decreased rotation blue. If refraction can alter rotation, so can reflection, just as the spin of a tennis ball is altered when it bounces. The type of surface determines what alteration occurs, and surfaces thus appear in different colors. For all its arbitrary and unconvincing elements, Descartes' treatment of colors was a major event in the history of optics. Not only did he abolish the distinction between real and apparent colors, placing them all on the same ground, but he was also responsible for incorporating phenomena of color into the science of optics. They have been there ever since.

They have not continued to be handled in Descartes' terms, however. His treatment of colors, for all its reaction against the Aristotelian conception of quality, merely transposed the peripatetic discussion of apparent colors into mechanical terms. It began with the assumption—another of the commonsensical assumptions of the scientific tradition, like the stability of the earth, which appeared so obvious that it could hardly be recognized as an assumption—that light in its primitive and natural state appears white. Colors arise when white light is modified by the medium through which it passes. In correlating red with a fast rotation and blue with a weak one, he even found a mechanical equivalent of the strong and weak colors of traditional theories. Throughout the 17th century, mechanical philosophers tended, not to challenge theories, but to imagine mechanisms to account for them. So it was in Descartes' discussion of colors, and so it was in the writings of Grimaldi, Hooke, and Boyle, who followed him. Although they altered the details of the mechanism, they did not think to challenge the assumption of modification.

That challenge was the work of an undergraduate in Cambridge University named Isaac Newton (1642–1727). In considering the colored fringes on bodies observed through a prism, Newton proposed a new approach to colors. Perhaps the rays that cause the sensations of different colors differ inherently from each other and are refracted at different angles. A prism then would cause colors to appear by separating the rays instead of modifying them. To test the idea, Newton observed a thread, one half of which was red and the other blue, through a prism; the two ends appeared to be disjoined. The new idea, confirmed by the experiment, was destined to turn the explanation of colors upside down—or perhaps right side up—placing it on the foundation it has occupied ever since. Newton's *Opticks,* not published until 1704, forty years after the initial insight, constituted a lengthy elaboration of the original insight.

If the new theory were seriously to be advanced, it would require a far more extensive experimental demonstration. Newton chose the prism as the instrument to provide it. His basic experiment modified the earlier one by Descartes. A prism projected the spectrum of a narrow beam of light, not onto a screen immediately behind the prism, but onto the wall across the room, a distance great enough to allow the rays to separate. (See Fig. 3.5.) The spectrum painted itself on the wall some five times as long as it was wide, whereas a round spot should have appeared if all rays are equally refrangible. The theory of modification

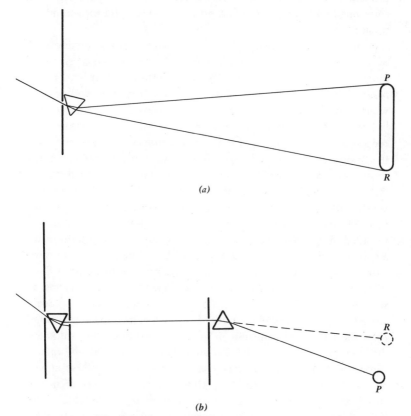

(a)

(b)

Figure 3.5.   (a) Prismatic projection. (b) The experimentum crucis.

had a possible explanation of the elongated spectrum. Already it held that the colors of the spectrum are modifications of white light produced by the prism; why may the dispersion of the beam not be another modification? To answer the objection, Newton devised what he called the *experimentum crucis*. (See Fig. 3.5.) Behind the prism, he set up a board with a small hole so placed that by rotating the prism slightly on its axis he could project different parts of the spectrum through the hole. Halfway across the room another board with a hole allowed the beam to pass. Since the two boards were fixed in place, they defined a constant path for the beam and a constant angle of incidence on a second prism set in a fixed position behind the second board. When the red end of the spectrum was projected through the holes and onto the second prism, it was refracted there at an angle corresponding to its refraction by the first prism; blue was refracted at a greater angle, again corresponding to that of the first prism. In no case did the second prism cause further dispersion.

White light, Newton concluded, is a heterogeneous mixture of rays which differ both in the sensations of color they arouse when they impinge on the eye, and in the degree of refraction they undergo in a prism. The phenomena of colors arise, not from the modification of white light, but from the separation of the rays that compose it.

Even with the addition of such things as rainbows, prismatic spectra constitute a small fraction of the color phenomena in the world. Before the theory could claim to be complete, Newton would have to extend it to include the colors of bodies; that is, he would have to demonstrate that reflection also can separate the mixture into its components. He developed his argument in a paper which he sent to the Royal Society in 1675. The foundation of the investigation was Hooke's observation, recorded in his *Micrographia* (1665), that thin plates or films of transparent materials such as mica appear to be colored, the colors varying with the thickness. Hooke could not imagine how to measure films so thin. Newton could. By pressing a lens of known curvature against a flat sheet of glass, he created a thin film of air between the two. (See Fig. 3.6.) In the film appeared a series of colored rings ("Newton's rings"), from the measured diameters of which the corresponding thickness of the film could be calculated. Newton demonstrated that if, in monochromatic light, a ring appears at a thickness x, then others appear at 3x, 5x, 7x, and so on indefinitely, while dark rings between them appear at 2x, 4x, 6x, and so on. The measurements hold for reflected light; if the film is viewed from the other side, the positions of bright and

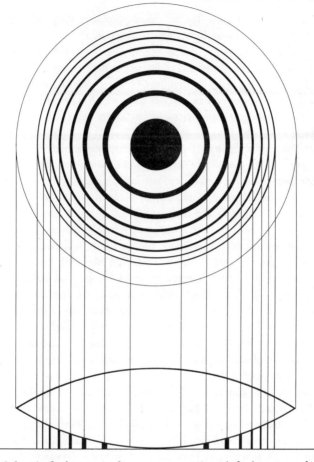

Figure 3.6. At the bottom is shown a cross section of the lens pressed against the flat sheet of glass, creating a thin film of air between them. At the top is shown the pattern of dark and bright circles observed in monochromatic light reflected from the film. A dark circle appears wherever the light is transmitted, so that none is reflected to the eye.

dark rings exactly reverse. That is to say, the rings appear because light is either reflected or transmitted by a given thickness of a transparent film, and the effective thicknesses are periodic. Moreover, the same thickness which reflects purple does not reflect red. Newton measured the thicknesses corresponding to the different colors of the spectrum with great precision, creating the factual basis on which later investigations of what we now call interference phenomena built. To him the significance of the results lay in the explanation they offered of the colors of bodies. Rays of light, which differ in the colors they exhibit and in their refrangibility, differ also in their reflexibility. The thickness of film which reflects red does not reflect purple. But the mechanical philosophy tells us that bodies are composed of particles of given size and shape. Obviously, bodies which appear red are composed of (transparent) particles the thickness of which is proper for the reflection of red, and so on.

When Newton's theory of colors was first published, his contemporaries generally failed to understand it. For over two thousand years, since the beginning of systematic natural philosophy, white light had been considered simple and primitive. Newton proposed on the other hand that white light is a mixture of heterogeneous rays, each one of which causes the sensation of a distinct color. Not white light, the mixture, but its components constitute simple light. When the reversal of concepts had been digested, it was seen that his work embodied a significant contribution to the mechanical philosophy of nature, placing on an experimental foundation what Descartes had placed solely on a speculative one. Accepting the proposition that colors cannot be real qualities of bodies, but only sensations caused by light, Newton had incorporated the theory of colors fully into optics. He had destroyed the distinction of real and apparent colors, and traced all sensations of color to identical principles.

Newton was as convinced of the mechanical nature of light as Descartes had been. Descartes had argued that light is a continuous pressure in a transparent medium; others following him, such as Hooke, had modified the view to consider light as individual pulses transmitted through media. From these suggestions what we know as the wave conception of light developed. A radically different conception was consistent with the basic premises of the 17th century's natural philosophy, and Newton embraced it. Light consists of tiny corpuscles moving with immense velocity. Not only does the rectilinear propagation of light correspond to the inertial motion of bodies, but Newton was convinced that the invariable properties of rays, the refrangibility, the reflexibility, and the color each exhibits, properties he found no way to alter in individual rays, demand a material foundation. Thus he believed that corpuscles which cause the sensation of red are bigger than those which cause blue. During the 1670's, he developed an elaborate theory to explain optical phenomena in mechanical terms. He imagined that all space is filled with a subtle matter called the aether and that variations in the aether's density cause corpuscles of light travelling through it to change direction. In these terms he explained reflection, refraction, and inflection (or diffraction as we call it—the bending of rays as they pass near bodies under certain conditions), and by attributing periodic vibrations to the aether, he even explained the phenomena of Newton's rings.

Newton's "Hypothesis of Light," in which he expounded his theory, was a typical product of 17th century mechanical philosophy. About four years after he composed it, he ceased to believe in the existence of an aether. When he published the *Opticks* early in the 18th century, he used attractions between particles to explain all the phenomena he had traced earlier to variations in the aether's density. All the phenomena, that is, but one—he could not explain the periodic phenomena of thin films. That the periodic phenomena exist was beyond doubt as his experiments proved, but the aethereal vibrations used to explain them were no longer possible. Hence Newton's *Opticks* appeared with a most peculiar passage about "fits of easy transmission" and "fits of easy reflection" in which he announced as facts what were inexplicable in his theory.

Newton's theory was not alone in this respect. There was no conception of light in the 17th century which was able to account for periodic phenomena, not even the so-called wave conceptions. If Newton was the leading exponent of the corpuscular theory, the Dutch scientist, Christiaan Huygens (1629–1695), played the corresponding role for the wave theory. Against the corpuscular view Huygens felt there were overwhelming objections. Rays of light can cross without interfering with each other, but streams of particles could not avoid interfering. Moreover, light spreads out through an immense sphere around a source; if the sun, for example, were continually to emit particles to fill the sphere it illuminates, its substance would waste away, and its size would be seen to diminish. Light, then, cannot be corpuscular. Since light is a mechanical phenomenon, it must be a motion transmitted through a medium.

"It is impossible to doubt that light consists in the motion of some sort of matter. For if we consider its production, we note that here on the earth it is caused principally by fire and flame which undoubtedly contain particles in rapid motion since they dissolve and melt a number of other bodies that are very solid; or if we consider its effects we see that when light is focused, as by concave mirrors, it has the power to burn like fire, that is, it separates the parts of bodies, which surely indicates motion, at least in the true philosophy in which the causes of all natural effects are conceived in mechanical terms. Which must be done in my opinion, or we must give up all hope of ever understanding anything in physics."

When a stone falls into a pool of water, it sets up waves which spread out from their center over the entire pool. The water itself does not flow away from the center, but the disturbance does move outward, passing from one particle of water to the next. Huygens' great contribution to optics was to demonstrate how a similar mode of propagation through a medium composed of hard particles (that is, through an aether) is compatible with the rectilinear propagation of light. The essential concept was that of the wave front. When a disturbance, arising ultimately from the rapid motion of the particles of a luminous body, is propagated through an aether, each particle of the aether in its turn becomes the center of a tiny wavelet spreading out from it as a center. (See Fig. 3.7.) Each of the wavelets alone is too weak to be perceived as light; only where a number combine to reinforce one another is the motion intense enough to constitute light. Huygens called the location of the reinforcements the wave front, and he showed that the wave front spreading out from a luminous point will be a sphere of which the point is the center. In reality, the wave front is constituted from an infinite number of wavelets reinforcing each other, but the result is identical to the spherical surface spreading out from the point.

The principal objection to the wave theory held that luminous waves would spread into the shadow behind an obstacle just as waves on a surface of water do. By applying the concept of the wave front, Huygens demolished the objection. Each wavelet is propagated into the shadow, but in this direction no wave front, where the wavelets reinforce each other, is formed. An effective wave front forms only along straight lines emanating from the source, and the wave conception of light yields rectilinear propagation as well as the corpuscular. Moreover, Huygens developed the demonstrations still found in elementary texts whereby his concept of the wave front also yields the laws of reflection and refraction.

The wave conception of Huygens and the corpuscular conception of Newton both accounted for much the same range of basic optical phenomena. Newton was able to incorporate the heterogeneity of light into the corpuscular theory; Huygens never succeeded in accounting for

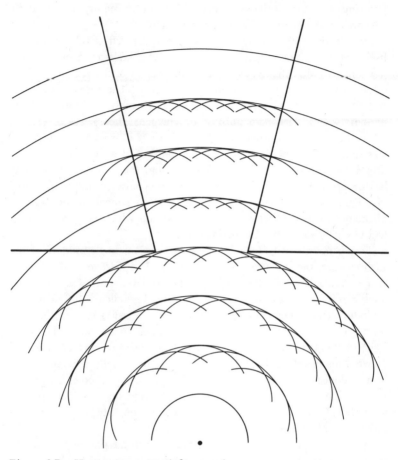

Figure 3.7. Huygens' concept of the wave front.

colors on his terms, although Malebranche suggested soon thereafter that each color represents a different frequency. The two theories differed in the relative velocities that they assigned to light in different media. For Newton, it was necessary that light move faster in media in which it is refracted toward the normal (such as glass or water when light enters it from air). For Huygens, it was just the contrary; light must travel more slowly in such media. In both cases, the necessity was dictated, not by arbitrary opinions, but by the need to make refraction follow the law of sines. Thus a crucial experiment to decide between the two theories existed; and in the middle of the 19th century, the experiment confirmed the wave theory. In the 17th century, it was utterly beyond the ability of optical experimentation.

Even in the 19th century, the wave theory had effectively established itself before the measurement of relative velocities was made; periodic phenomena rather than velocities bore the weight of the argument. Why could they not do the same in the 17th century? The answer lies in the fact that the so-called wave theory of the 17th century did not embody periodic waves. The main purpose of Huygens' theory was to explain the rectilinear propagation of light, and when he spoke of a wave he was thinking, not of a periodic fluctuation, but of a disturbance travelling across the surface of a pond away from the point where a stone has dropped. He specifically denied that the pulses could be periodic. His *Traité de la lumière** (1690) did not even mention the periodic phenomena which Newton had discovered and which Huygens himself had observed in identical experiments.

That is not all that Huygens failed to mention. In the middle of the 17th century, the Italian scientist, Grimaldi, had discovered diffraction. A few years later, Bartholinus had discovered a phenomenon called double refraction, which involved polarity. Periodicity, diffraction (a periodic phenomenon, though it was not recognized as such in the 17th century) and polarity became the foundations of the wave theory of light in the 19th century. Huygens did not mention any of them. If Newton discussed all three, his position was not wholly different. After he abandoned belief in a vibrating aether, periodic phenomena appeared inexplicable to him. His explanation of diffraction was incompatible with the phenomena, and his explanation of polarity was difficult to reconcile with the rest of his theory of light.

By the end of the 17th century, the mechanical philosophy, which encouraged optics early in the century, and which furnished the idiom in which all students of optics, including Newton and Huygens, discussed the science, had become an obstacle to its further progress. Experimentation had discovered three properties or phenomena which were frankly unintelligible to either of the mechanical models in use. Until a much subtler wave mechanics was developed, emphasizing not the mechanical medium, but the wave motion itself, optics stagnated for a century.

* *Treatise on Light.*

CHAPTER IV

Mechanical Chemistry

THE CHEMISTRY which the 17th century inherited was cast in a mold so different from modern chemistry that a reader from the 20th century must transport himself by the imagination into an intellectual climate utterly unlike his own. In sciences such as astronomy and mechanics, the basic problems of the earlier traditions look familiar even if the conceptions with which they were handled seem strange. In the case of chemistry, a major effort is required even to recognize the problems.

Part of the difficulty arises from the idea of mixed bodies. The chemist considered that all the bodies (or materials) he handled were mixed bodies. It was universally agreed that a limited number of elements or principles unite in various proportions to compose the materials found on the surface of the earth. The word "principle" is less apt to mislead the modern reader than the word "element," because element has been employed by later chemistry to signify a concept which has little or nothing in common with the earlier one. The number of principles (or elements) varied with the system; usually there were three, four or five—always a number, not just smaller than, but of a different order from the number of elements of which we are accustomed to think. These principles were the universal ingredients of all the bodies found on the surface of the earth. All the principles or elements entered in some proportion into the composition of every mixed body. From the bewildering variety of materials found in nature it appeared to follow that the possible proportions were infinite in number. Hence another significant divergence from later chemistry—if the number of principles was radically smaller, the number of chemicals was infinitely larger. Instead of a discrete number of compounds, the chemist of 1600 thought in terms of a continuous spectrum of possible proportions. One saltpetre was not identical to another, and the chemist needed to specify the origin of his material. Undoubtedly the presence of impurities dictated such a notion, but the very idea of mixed bodies in a continuous spectrum of proportions made it virtually impossible to distinguish a given chemical from impurities. Without any criterion to identify the chemical, how was one to recognize the impurities? In the case of the precious metals, considerations of the most practical kind had long since established criteria of purity, but chemists continued to discuss different "golds" and "silvers." They continued also to think of metals as mixed bodies. To understand the practical problems facing the 17th century chemist, nothing is more instructive than two essays by Robert Boyle on the unsuccessfulness of experiments, written some fifty years after the period being discussed. Still groping toward a set of adequate concepts, chemistry at the beginning of the century was simply overwhelmed by the variety of phenomena.

The work of chemistry was analysis. By various means, almost all involving fire, the chemist separated mixed bodies into their elements or principles. As an historian of the 20th century has pointed out, "analysis" had a meaning different from the one we are accustomed to, just as the word "element" had. Since our elements are concrete identifiable substances, we expect to isolate them by analysis. In 1600, the analysis of the chemist was rational rather than real. He intended his manipulations to reveal the composition of mixed bodies, but he did not expect to isolate them as concrete substances he could handle. The very terms in which the principles were conceived denied the possibility of their isolated existence.

Another feature of chemistry at the beginning of the 17th century was its relation to the broader discipline of natural philosophy. Chemistry as a distinct science scarcely existed. In so far as chemistry was a distinct enterprise, it was not generally considered a science. In so far as it was part of science, on the other hand, chemistry was not a distinct enterprise. Chemists themselves looked upon their subject as an art in the service of medicine; their efforts were devoted to the compounding of medicines. Scientists looked upon them with scorn, referring to them, in a phrase that can hardly be misunderstood, as "sooty empirics." In the work of Paracelsus (1493?–1541), chemistry had reached one of its most developed forms. It is difficult to read Paracelsus without concluding that he used chemical phenomena primarily to illustrate a philosophy basically concerned with religious questions. Although his concepts and his theories were to exercise an influence on chemistry which endured through the 17th century, they were not formulated initially to handle chemical phenomena. Quite the contrary, phenomena were pressed into the mold that the concepts furnished.

Paracelsus was the leading influence on chemistry at the beginning of the 17th century. Around his teachings, a school known as "iatrochemists" or "spagyrists" had formed, a school which saw chemistry as the servant of medicine. The books that they published—and there was a tradition of iatrochemical texts that extended unbroken through the 17th century—consisted mostly of medical recipes introduced by a mere modicum of theory.

The theory was based on the Paracelsian teaching of three principles —salt, sulfur, and mercury. From the three principles as components, all mixed bodies are compounded. To Paracelsus, salt, sulfur, and mercury represented body, soul, and spirit, three metaphysical constituents of all existent bodies. Though less inclined to metaphysical speculation, the iatrochemists never wholly divested the principles of their original character. Jean Beguin (c. 1550–c. 1620), a leading iatrochemist of the early 17th century, defined mercury as an acid liquid, permeable, penetrating, and aethereal. To it are due the sense and movement of bodies, their forces, and their colors. Sulfur is a gentle balm, oily and viscous, which conserves the natural heat of bodies and renders them inflammable. It is the instrument of vegetation, growth, and transmutation, and the source of odors. It has the power to reconcile contraries, joining the liquidity of mercury to the solidity of salt. The last of the principles, salt, is dry and briny, the source of the solidity of bodies. While preserving the characteristics of Paracelsus' notion of body, soul, and spirit, Beguin's definition of the principles reminds us more clearly of Aristotle's elements. Salt corresponds to earth, sulfur to fire, and mercury to water. Like the Aristotelian elements, the three principles were conceived in qualitative terms; they were the material carriers of specific qualities. Thus a body was solid because it contained a high proportion of salt, or inflammable because of its high proportion of sulfur. In its acceptance of a qualitative conception of nature, iatrochemistry stood in conflict with the quantitative outlook that increasingly dominated the physical sciences in the 17th century.

The Paracelsian tradition of active principles, an aspect of Renaissance Naturalism, was equally at odds with the mechanical philosophy. Among the iatrochemists, it was common, though not universal, to consider the three Paracelsian principles as active and to admit beside them two passive principles, water and earth. Helmont, perhaps the last great Paracelsian, insisted that an active principle, analogous to the Paracelsian mercury, is the crucial constituent of every body. The view was diametrically opposed to the conception of bodies held by the mechanical philosophy.

Another strand of the tradition behind 17th century chemistry, alchemy, further emphasizes the dichotomy between the prevailing outlook in chemistry and the increasingly dominant mechanical philosophy. Although the iatrochemists of the 17th century were a sober and prosaic lot for the most part, and hardly up to the flights of fancy demanded of an alchemist, Paracelsus himself had been closely allied to it, and the alchemical conception of nature was wholly in accord with his. Alchemy looked upon metals as fundamentally identical with each other, differing only in degree of maturity. Gold, of course, is the most perfect, as its resistance to decay and corruption indicates. Silver stands next, with the others ranged below it. When the natural process which produces metal in the earth is carried to completion, it generates gold. When it is interrupted, when it aborts, it stops short at one of the baser metals. The work of alchemy, baldly stated, was to grow gold, using art to carry to fruition the natural process whereby gold is produced in the earth. Alchemy expressed the organic conception of nature in its most vivid terms. Its vocabulary was filled with words unmistakable in their connotation—fermentation, vegetation, digestion, generation, maturation. In its lengthy processes, it tended to employ organic heats; for example materials were buried in manure piles for periods of gestation. During the 17th century, the waning influence of alchemy made common cause with iatrochemistry in supporting a conception of metals which considered them as organic substances growing in the earth and as mixed bodies composed from the chemical principles.

In the middle of the 17th century, every major aspect of the chemical tradition expressed a view of nature profoundly opposed to that which was becoming dominant elsewhere in physical science. Descartes' implicit attitude toward chemistry is indicative of the divergent outlooks. Whereas he devoted chapters to the discussion of topics such as magnetism and light, paragraphs were the most he had for chemical questions. The fact that a few chemical questions did make their way into his work was itself indicative of the future. For chemistry was immediately and inescapably relevant to the mechanical philosophy of nature. If the properties of bodies are appearances caused by the particles of which they are composed, chemistry had much to say that a mechanical philosophy of nature could not ignore. The story of chemistry in the second half of the century is the story of its conversion to the mechanical philosophy. Perhaps one should say rather its subjection to the mechanical philosophy, since the growing role of mechanisms in chemical literature appears less to have sprung from the phenomena than to have been imposed upon them by external considerations. Be that as it may, from the vantage point of the 20th century it appears impossible that chemistry could have continued untouched by the influence of the prevailing philosophy of nature. In fact it did not. If most of the leading chemists of the first half of the century were Paracelsians, the leaders of the second half were nearly all mechanists.

In assessing the shift in chemical thought that occurred, we must recall the internal history of iatrochemistry as well. When the Paracelsian principles were originally formulated, they applied to a rather limited body of chemical data, much of it organic. We still use the Paracelsian word "spirits" to describe the products of certain organic distillations, which contributed a significant proportion of the corpus of chemical information. In the 17th century, the corpus of chemical knowledge expanded greatly. The bulk of the new information, moreover, concerned inorganic chemistry, and considerable effort was required to force much of it into the categories of Paracelsian theory. The apparent collapse of iatrochemistry in the second half of the century must be judged against the background of the growing body of chemical knowledge. The iatrochemical tradition catalogued the reactions and the preparations of chemicals. It failed utterly to organize the facts into a coherent and useful body of theory and its own failure contributed to the triumph of mechanical chemistry. The principal question confronting chemistry in the latter half of the century was whether the mechanical philosophy could do what iatrochemistry had not.

Nicolas Lemery (1645–1715) was the leading chemist in France during the second half of the century. His *Cours de chimie** appeared initially in 1675 and through its many editions and translations exercised an extensive influence on chemistry. The tone of the work is adequately represented by the discussion of the means by which spirit of sea salt (hydrochloric acid, HCl) precipitates metals that *aqua fortis* (nitric acid, HNO_3) has dissolved. The bulkier pointed particles of the spirit of sea salt jostle and shake those of the *aqua fortis* until the metal they hold in solution is dropped. Some chemists, he added, explain the reac-

---

* *A Course of Chemistry.*

tion by the conjunction of the acidity of the spirit with the volatile and sulfuric alkali of the *aqua fortis,* which constrains the latter to abandon the metal.

"But this is the way to explicate, as they say, one obscure thing by another that is much more obscure; for what likelihood is there that the volatile spirit of *Aqua fortis* is an alkali? and pray how comes it to remain in so great a motion with the fixed acid Spirit of this same water without destroying or losing its nature, this is a thing that can never be conceived very easily. But furthermore supposing this Spirit were an alkali, it would be still necessary to explicate mechanically, for what reason this alkali does quit the body of the metal to betake itself to the *Spirit of Salt;* for to say meerly that by the conjunction of these two Spirits the *Aqua fortis* is compelled to abandon the metal that it had dissolved, is nothing at all to the clearing of the question, unless a man will needs give an intelligence to these Spirits. Wherefore we must still have recourse to the agitation and jostles, for the true reason."

Because of the significant role of acid reactions in his chemistry, passages similar to the one above appear frequently in Lemery's text. Acids are composed of pointed particles—he often called them "acid points." They are very light as well, so that the acid points can hold up metallic particles they have impaled just as wood can make metal attached to it float. Incidentally, the image also explained why a given quantity of acid dissolves only so much metal; once every point has engaged a particle of metal, the acid can dissolve no more. Why is it then that solvents quit bodies they hold in solution and betake themselves to another—that is, for example, why does an alkaline salt precipitate gold from *aqua regia* (a mixture of nitric acid and hydrochloric acids—HNO_3 and HCl)? The question is one of the most difficult in natural philosophy, Lemery agreed, but apparently not so difficult that his mechanical philosophy could not answer it. When you add to the solution some material adapted by the figure and motion of its particles to engage and break the acid points which impale the particles of gold, it must precipitate the gold. It happens that volatile alkaline spirits are impregnated with "very active Salts" which move and shake the bodies they meet so violently that the points are broken. The points, though broken, continue to be sharp enough and active enough to impale the particles of the salt, however, and thus dissolve them with heat and boiling.

The nature of a thing, Lemery asserted, cannot be better explained "than by admitting to its parts such figures as are answerable to the effects it produces." The phrase is significant. It indicates the ultimate concern of Lemery's mechanical chemistry—not the formulation of chemical theory, but the explanation of observed properties. The properties of acids suggested sharp pointed particles. Corrosive salts formed from acids, such as the vitriol of copper, derive their power to corrode from the acid particles sticking in them, or better out of them, like so many unsheathed knives which cut and shred the materials they meet. As an instrument of explanation, the mechanical philosophy did not in itself offer a chemical theory. On the contrary, it was potentially adaptable to almost any theory. The particles of given shapes that Lemery and others discussed were not observed in any sense whatever; they were inferred from the observed properties, and in practice it was possible to imagine particles of any shape and motion that were required for the phenomena in question.

In Lemery's case, the mechanical philosophy was employed to explain a modified version of iatrochemistry. He virtually ruled mercury or spirit out of chemistry; all the materials ordinarily referred to as spirits are really volatile salts. There is a universal spirit (it is difficult to know if he meant it to be material or immaterial) which is the ultimate cause of all specific substances, but Lemery considered it too metaphysical and abstract to be included in a treatise on chemistry. Sulfur, or oil as he preferred to call it, consists of pliable, ramous particles which entangle other particles and themselves and thus reveal the familiar viscosity of oil and grease. Oil continued to be the principle of inflammability for Lemery, but he also believed in the existence of tiny round igneous particles, so that his treatment of combustion is utterly incomprehensible. In fact, Lemery did not devote much space to oil in his work, virtually all of which is concerned with the third Paracelsian principle, salt. There is one salt in nature, Lemery argued, an acid salt formed by the solidification of an acid liquor in the veins of the earth. All other salts are formed from it. Alkali salts have no natural existence in mixed bodies but are created in them by the chemical operations which appear only to separate them. Lemery insisted on this principle as the key to the interpretation of chemical phenomena. Needless to say, it was his discovery. He did not make extensive use of it, however, whereas the existence of alkalis, whatever their origin, was of the most central importance to his chemistry. If there were any organizing scheme in Lemery's Cours de chimie, it was the apparent attempt to reduce most reactions to the neutralization of an acid by an alkali. Since van Helmont first described such a neutralization, its importance in Lemery's chemistry is another facet of his relation to the Paracelsian school.

Acids are composed of sharp pointed particles, pins. Alkalis are composed of porous particles in which the points can stick, pin cushions. The two mix with ebullition, a fact acknowledged but not explained in any satisfactory terms. The pin sticks in the pin cushion and is neutralized; the pin cushion has its pores filled by pins and is neutralized. In various forms, which is to say sharper points or blunter points, larger pores or smaller pores, Lemery repeated the image endlessly throughout his work. Product of the iatrochemical tradition that he was, he was concerned with the medical applications of his work and applied his basic scheme to that field as well. Diseases are acid infections carried by the air and as such basically like poisons. The whole of medicine consists of their neutralization by alkalis. As he said by way of warning in the chapter on corrosive sublimate, one must be sure he knows the nature of poisons before he prescribes an antidote. Since the whole of Lemery's work stressed the similarity of alkalis above their difference, he left the wondering doctors without much useful guidance.

Perhaps the most pervasive influence of iatrochemistry on Lemery is found in his tendency to think in terms of a few broad classes of substances. One basic shape accounted for the acids and another for the alkalis, and whatever the dictates of experience, he constantly implied the ultimate identity of all acids and of all alkalis, and indeed perhaps of all substances. Chemistry in his treatment was devoted, not to the separation and combination of enduring substances, but to the molding of malleable particles into a few general shapes. The earlier conviction of a continuous spectrum of mixed bodies found its counterpart in the continuous variation of shapes implicit in Lemery's discussion. Acids differ according to the sharpness of their points. Is each point a distinct and immutable shape corresponding to a specific acid? Apparently not, because he argued that the sharper particles are the products of a longer fermentation in the earth whereby they are beaten to finer points. So also, when *mercurius dulcis* (calomel, Hg_2Cl_2) is made from corrosive sublimate ($HgCl_2$), the material must be sublimed three times to blunt the acid points. If it is sublimed only twice, the points will remain too sharp and their purgative power too strong. On the other hand, if it is sublimed five times, the purgative power is wholly destroyed, and it becomes merely sudorific. One of the fundamental propositions of the mechanical philosophy was the homogeneity of matter which is differentiated only by the shapes, sizes, and motions of its particles. The means to translate the idea of a continuous spectrum of proportions into mechanical terms were ready to hand, and Lemery seized them without pausing to reflect.

How far his mechanical prepossessions could obscure the meaning of important observations is suggested by his treatment of the decomposition of mercuric nitrate $(Hg(NO_3)_2)$ into mercuric oxide (Hg_2O_2) by means of heat. When the white crystals formed from the evaporation of a solution of mercury in spirit of nitre (HNO_3) are heated, they decrease in weight and turn red—obviously because the edges of the acid points are struck off. Another red precipitate can be formed merely by calcining mercury. In this case, igneous particles enter the pores of the mercury, giving its particles a new disposition and motion whence the red precipitate results. It did not occur to Lemery that the two red compounds could be identical. His work contains an extensive list of specific chemicals with instructions on how to prepare them. The very fact that definite substances identified by invariant properties exist seems to stand in direct contradiction to his discussions of the plasticity of matter. All the preparations of mercury, he said, "are nothing but so many different shapes of *Mercury* made by acid Spirits, which according to the different adhesions, do cause different effects." But as he also noted, the original mercury can be revived from its compounds. That fact appears to us to contradict the infinite malleability of matter, but Lemery made no effort to resolve the dilemma.

In the case of Lemery, the mechanical philosophy served neither to criticize the chemical theory he received nor to suggest an alternative. By focusing his attention on the imagined shapes of particles presumed to explain the properties of substances, it tended to divert his attention from the reconsideration of the extensive body of data he possessed. Like his fellow mechanical chemists, he seemed possessed by a mania to explain every property and every phenomenon. If anything, the mechanical philosophy operated to perpetuate the traditional body of theory by encouraging chemists to imagine invisible mechanisms which appeared to bring it into harmony with the accepted philosophy of nature. Although iatrochemistry had failed as an independent theory, in a disguised form it influenced the shape of mechanical chemistry.

The work of John Mayow (1640–1679), an English doctor and chemist, further illustrates how easily the mechanical philosophy could function to sustain a traditional point of view in chemistry. Mayow was one of a number of experimenters interested in the analogies of respiration and combustion. It was known that when a candle is burned in a closed container over water, water rises in the container and the volume of air decreases as the candle burns out. (See Fig. 4.1.) Experiments now determined that the same phenomenon (in about the same quantity) occurs when a small animal expires in a similar closed container. Not only did the reduction in volume suggest that something is removed from the air, but experiments with the air pump supported the conclusion by showing that the presence of air is necessary for both combustion and life. To these known experiments Mayow added another. He enclosed a small animal in a jar together with some combustible material that could be ignited by a burning glass. When the animal had expired, the combustible material could not be set on fire; therefore both respiration and combustion require the same substance in the air. Mayow referred to it as the nitro-aerial spirit, a name deriving from nitre (saltpetre) and expressing the fact that materials made from nitre, such as gunpowder, which contain their own nitro-aerial spirit, can burn without the presence of air.

After the work of Lavoisier in the 18th century interpreting the role of oxygen in combustion and respiration, Mayow was hailed as a precursor to its discovery. In fact, however, his work can be understood far

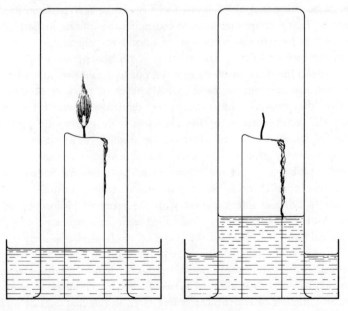

Figure 4.1. Mayow's experiment.

better in relation to the Paracelsian tradition, from which the very name nitro-aerial spirit came. The clue to understanding him is found in his interpretation of the decrease in volume of the air. According to our chemistry, oxygen combines with carbon in both respiration and combustion to form carbon dioxide which dissolves in the water. In contrast, Mayow argued that the elasticity of the air is decreased. He looked upon the nitro-aerial spirit, not as a separate gas which forms part of the air, but as the cause of the elasticity of air. In mechanical terms, he spoke of nitro-aerial particles wedged in the particles of air so as to make them elastic. Perhaps a suitable analogy (which he did not use) might be to picture particles of air as empty tubes which are given firmness and elasticity by pieces of wire (nitro-aerial particles) extending through them. Mayow's nitro-aerial particle was nothing more than a Paracelsian active principle tricked out in mechanical attire. It causes the elasticity of air. It is responsible for combustion. Separated from the air in respiration, it maintains animal life. When it ferments with saline-sulfurous particles in the blood, it produces animal heat, and when it effervesces again with other saline-sulfurous particles supplied through the nerves, it causes muscles to contract and is thus the source of animal motion. Nitro-aerial spirit is also responsible for vegetable life. One begins here to suspect the anomalous role that saltpetre (potassium nitrate) played in Mayow's theory. An ingredient of gunpowder, saltpetre was dug from earth plentifully supplied with manure; hence it was associated with fertilizer. In addition, spirit of nitre (nitric acid) could be made from it. At least three different elements as we know them were jumbled together in Mayow's nitro-aerial spirit. As the agent of combustion and animal life, it referred to oxygen; as the agent of vegetable life, to nitrogen; as the acidic spirit, to the hydrogen ion. Chemistry in the 17th century was not sophisticated enough to make these distinctions of course. As far as Mayow was concerned, nitro-aerial particles were a mechanical rendition of the active spirit of Paracelsian chemistry.

Easily the most important mechanical chemist was Robert Boyle. Boyle liked to pose as a Baconian empiricist, engaged in collecting a natural history, unencumbered by theoretical prepossessions. Several times, in prefaces, he asserted that he had forborne to read Descartes and Gassendi lest he be seduced by their systems. Nothing could be more misleading. From the very beginning of his scientific career, Boyle was committed to the mechanical philosophy, and the entire corpus of his scientific writing can be interpreted as a running exposition of it. Scion of an enormously wealthy family (the father of chemistry and the brother of the Earl of Cork, according to one wit), Boyle had the means to pursue what studies he chose. During the 1650s, he settled in Oxford, became the associate of the group who later formed the nucleus of the Royal Society, and chose chemistry. His friends were visibly upset; chemistry was not a science. Boyle thought otherwise—it appeared to him that chemistry was uniquely placed to provide the mechanical philosophy of nature with an experimentally based theory of matter. To this goal he devoted his scientific career. Early in the career, he projected a great work on the corpuscular theory of matter. The work was never completed, but a large number of separate treatises which may be thought of as contributions to the master work were. One theme dominates them—the argument that chemical reactions are merely the reshuffling of particles and that all chemical properties are the products of particles of matter in motion.

In the *Sceptical Chymist* (1661), one of his early works, Boyle defined an element—perhaps his best known statement. "I mean by elements, certain primitive and simple, or perfectly unmingled bodies; which not being made of any other bodies, or of one another, are the ingredients, of which all those called perfectly mixt bodies are immediately compounded, and into which they are ultimately resolved." The words have been frequently quoted and as frequently misunderstood. Far from proposing a new conception of an element, they express the traditional conception of elements or principles as the components of mixed bodies, a concept which Boyle then proceeded to reject in the following clause of the same sentence. In its place, he substituted his own version of the mechanical philosophy. Matter consists of a multitude of uniform tiny corpuscles, which unite together to form larger particles, which in turn constitute the materials and bodies chemistry handles. All the differences we observe among bodies must derive from differences in the shape and motion of the secondary clusters, the effective particles from which bodies are composed. Exactly this point, repeated endlessly in Boyle's endless sentences, constitutes the basic theme of his work.

Boyle was particularly concerned to apply the mechanical conception to chemical reactions. One of his most revealing essays carried the title, "The Redintegration of Saltpetre." He described an experiment in which saltpetre (KNO_3) was separated into a volatile spirit (HNO_3) and a fixed salt (K_2CO_3, the carbon derived from charcoal that he used in the experiment). Although he was not able to collect the spirit of nitre that was driven off, he could measure it from the loss in weight of the original saltpetre. When a quantity of spirit of nitre nearly equal to the amount driven off was dropped onto the fixed salt, saltpetre equal in weight to the original sample was reconstituted. It was not necessary to start initially with saltpetre. Since saltpetre is the combination of a specific acid spirit with a specific alkalisate salt, a saltpetre differing in no way from the natural substance can be compounded artificially. The two components, moreover, have qualities contrary to each other and to those of saltpetre, and it is difficult to see how the properties of saltpetre can derive from its components as the traditional chemistry wished to claim. The properties of saltpetre derive rather from the shape of its particles, which are composed from the particles of the two substances that compose saltpetre. The development of a satisfactory chemical theory as such was not Boyle's goal. Chemistry represented to him a means to demonstrate the validity of the mechanical philosophy of nature.

As with Lemery and Mayow, his chemistry retained a considerable deposit from the Paracelsian tradition behind its mechanical facade. When he set out to analyze saltpetre into its components, he turned to fire as the agent of decomposition without a moment's hesitation, and the two products presented themselves to him as a volatile spirit and a fixed salt. He agreed that the acid spirit embodied the active ingredient of saltpetre; the notion of active substances almost forced itself on the consciousness of a chemist, and the mechanical analogy of motion appeared so obvious that he did not pause to ask if "active" substances are in fact compatible with a mechanical conception of nature. Boyle agreed that metals grow in the earth, and that "seminal principles" (van Helmont's term) produce them. Helmont's experiment with the tree, leading to the conclusion that all things are made from water, agreed with the mechanical philosophy's premise that all bodies are formed from a uniform matter differentiated only by the shape and motion of its particles. Boyle cited the experiment continually, and he performed it twice himself.

Hence the mechanical conception suggested the universal mutability of substances whereby one might be changed into another.

"I would not say, that any thing can immediately be made of every thing, as a gold ring of a wedge of gold, or oil, or fire of water; yet since bodies, having but one common matter, can be differenced but by accidents, which seem all of them to be the effects and consequents of local motion, I see not, why it should be absurd to think that (at least among

inanimate bodies) by the intervention of some very small addition or
subtraction of matter, (which yet in most cases will scarce be needed,)
and of an orderly series of alterations, disposing by degrees the matter
to be transmuted, almost of any thing, may at length be made any
thing."

Almost any thing can be made from any thing—the mechanical philosophy provided such a ready image into which the old belief in transmutation could be translated that Boyle never questioned it. Experiments
did suggest that some substances are rather durable. Silver or mercury
could be put through a considerable series of reactions yielding one
substance after another, from which the original silver or mercury could
be reclaimed. Boyle's conception of matter offered a rationale to explain
the facts. The particles of silver and mercury are secondary concretions
of ultimate corpuscles which are very tightly bound together; the particles endure through the series of experiments intact, and the identical
metal exists in its compounds. The image seems to be pregnant with
the most fruitful possibilities. Not only a few metals but other substances as well—the two components of saltpetre for example—apparently consist of enduring particles, and we see Boyle standing on
the verge of a chemistry devoted to the combination and separation
of a finite number of distinct substances. Boyle lived in the 17th century, however, not in the 19th, and the possibilities we see were not
apparent to him. The experimental evidence he cited never led him to
question that metals are compounds, and once again his mechanical
image helped to confirm a traditional conviction. If the secondary concretions are relatively durable, they still remain secondary concretions
capable of being dissolved. Boyle continued to search for the means of
transmuting gold and to exchange secret recipes with other well known
alchemists such as John Locke and Isaac Newton. He continued to regard metals as mixed bodies, and substances such as water and alcohol
(spirit of wine) as more elementary.

Nevertheless, Boyle went further than any other chemist of his generation in questioning the existing structure of chemical theory. Lemery
simply plastered the existing theory with a thin layer of mechanical
explication. Whatever the elements of tradition that survived in his
chemistry, Boyle subjected both the Paracelsian principles and the Aristotelian elements to a searching critique in his *Sceptical Chymist*. Since
both elements and principles had been conceived as the material carriers
of qualities, they were bound to be objectionable to a thoughtful mech-

anist such as Boyle, but the gravamen to his critique rested on other grounds. The chemical doctrine held that analysis separates mixed bodies into the principles; Boyle employed chemical tests to prove that different bodies yield widely different substances in analysis, and that different modes of analysis separate the same substance into different components. Chemical identification tests were not original with Boyle, but he employed them more extensively and carried them to a higher level of efficacy than any chemist before him. Boxwood yields a spirit in distillation as other woods do; Boyle showed that it differs from ordinary spirit of wood. Spirit of boxwood dissolves corals and boils and hisses with salt of tartar, whereas ordinary spirit of wood does not dissolve coral and lies quietly with salt of tartar. Spirit of boxwood turns syrup of violets red; ordinary spirit of wood leaves it blue. Saltpetre analyzes into the acid spirit of nitre and the alkali salt called fixed nitre (potassium carbonate). Spirit of nitre dissolves many metals; fixed nitre precipitates them. Fixed nitre dissolves many unctuous and sulfurous bodies; spirit of nitre precipitates them. Spirit of nitre turns a scarlet tincture of brasil yellow; fixed nitre turns it red again. Saltpetre itself does not alter the color of the solution. The elements and principles of earlier chemistry had been identified primarily by physical properties—salt by solidity, for example, and spirit or mercury by volatility. Implicit in Boyle's use of the chemical tests was a radically new idea of a chemical substance as one which answers to a series of chemical identification tests.

"And indeed since to every determinate species of bodies there doth belong more than one quality, and for the most part a concurrence of many is so essential to that sort of bodies, that the want of any one of them is sufficient to exclude it from belonging to that species; there needs no more to discriminate sufficiently any one kind of bodies from all the bodies in the world, that are not of that kind."

When chemistry finally followed the implications of this conception, it established the foundations of modern chemical theory. It implies, not an infinite continuum of proportions or an infinite malleability of matter, but the existence of a discrete number of substances identified by a precise series of tests. Boyle was unable to commit himself to his own doctrine, for if he stated the concept above he also believed that any thing can be made out of any thing. Again his mechanical philosophy appears to have operated to thwart the most promising aspect of his chemistry. By supporting traditional concepts and lending them a bogus respectability, it encouraged him to go on working at transmutations despite the fact that a consistent application of his identification tests could not have failed to convince him that he at least could not make any thing from any thing.

During the 1660s and 1670s Isaac Newton, then a young professor at Cambridge University, poured over Boyle's writings and extracted from them materials that figured prominently in his speculations on the structure of matter. In 1706, the chemical speculations were published as a Query attached to the first Latin edition of the *Opticks,* what is now numbered Query 31 in English editions. It represents one of the highest levels that chemical thought reached in the 17th century. Newton's chemistry was yoked to a mechanical philosophy of nature quite as closely as Boyle's, although he differed from Boyle and from most of the 17th century in asserting the existence of forces between particles. Where Boyle saw in chemistry an instrument to demonstrate that all the phenomena of nature derive from particles of matter in motion, Newton saw in its phenomena proof that particles of matter attract and repel each other.

"When any metal is put into common water, the water cannot enter into its pores to act on it and dissolve it. Not that water consists of too gross parts for this purpose, but because it is unsociable to metal. For there is a certain secret principle in nature by which liquors are sociable to some things and unsociable to others. But a liquor which is of it self unsociable to a body may by the mixture of a convenient mediator be made sociable. And water by the mediation of saline spirits will mix with metal. Now when any metal is put in water impregnated with such spirits, as into Aqua fortis, Aqua Regis, spirit of Vitriol or the like, the particles of the spirits as they in floating in the water, strike on the metal, will by their sociableness enter into its pores and gather round its outside particles, and by advantage of the continual tremor the particles of the metal are in, hitch themselves in by degrees between those particles and the body and loosen them from it."

If a substance is added, such as salt of tartar (K_2CO_3), which is more sociable to the acid, the acid particles will gather round it and the metal will precipitate. Newton's explanation of reactions was no less speculative than Boyle's and the forces of attraction (what he called sociableness in the passage above) no more empirical than the shapes of Boyle's particles. For chemistry, however, they had the advantage of focusing attention on the most fruitful aspect of Boyle's work, the concept of a substance identified by specific chemical properties. For example, from Boyle he learned the series of displacement reactions partially indicated above—copper displaces silver from an acid solution, iron displaces copper, salt of tartar displaces iron. Newton's chemical writings were concerned, not with broad classes such as salts or spirits, but with specific chemicals and specific reactions. Perhaps his conviction as an atomist that the particles of substances are immutable rather than malleable encouraged this point of view. Certainly he thought that the particles of each substance had specific attractions and specific repulsions to other particles. Hence the attention of chemical experimentation should focus on the chemical properties of substances. Newton's 31st Query was the primary influence behind the study of affinities which played a leading role in chemistry early in the 18th century and helped to prepare the way for Lavoisier.

When the 17th century closed, the body of data at chemistry's command was immeasurably greater than it had been a century before. The intensive experimentation of a hundred years had not been without gain. We cannot, however, miss the fact that great progress had not been made in chemical theory. The mechanical philosophy, which dominated chemical thought in the second half of the century, offered only a language in which to describe reactions. Since there were no criteria by which to judge the superiority of one imagined mechanism over another, the mechanical philosophy itself dissolved into as many versions as there were chemists. In no area of science was the tendency to imagine invisible mechanisms carried to more absurd extremes. It is difficult to see that the mechanical philosophy contributed anything to the progress of chemistry as a science.

One thing mechanical chemistry did achieve. It ushered chemistry into the boundaries of natural science. When the century opened, chemistry was not generally considered to be part of natural science; at worst it was occult mystification; at best it was an art in the service of medicine. By the end of the century, chemists occupied honored positions in the scientific societies of Europe. There can be little doubt that mechanical chemistry played a major role in the change. By stating chemistry in terms acceptable to the scientific community, it made chemistry respectable as it had not been before. When Descartes composed his system of nature in the 1630s and 1640s, he nearly ignored chemical phenomena. In 1700, he would not have dared to do so.

CHAPTER V

Biology and the Mechanical Philosophy

THE RAPID ACCELERATION of scientific enquiry in the 17th century was not confined to the physical sciences. If in the end the proudest achievements were recorded in that area, nevertheless biology (though it was not yet given that name) received an immense investment of attention and witnessed considerable discoveries as well. The concept of a scientific revolution has validity for the organic sciences as well as the inorganic.

During the century, a flood of new information swept over the life sciences. Overseas exploration brought knowledge of a host of new plants and animals; the microscope revealed new realms of life; intensified anatomical research uncovered new information about what had been considered well known. Thomas Moffett's attempt to classify grasshoppers revealed the dangers in too much information.

"Some are green, some black, some blue. Some fly with one pair of wings, others with more; those that have no wings they leap, those that cannot either fly or leap, they walk; some have longer shanks, some shorter. Some there are that sing, others are silent. And as there are many kinds of them in nature, so their names are almost infinite, which through the neglect of Naturalists are grown out of use."

The deluge of new knowledge beyond the power of biology immediately to assimilate suggests a major difference from physics. The revolution in physical concepts was a matter primarily, not of new facts, but of new ways of looking at old facts. In contrast, biological science witnessed for the most part an enormous expansion of its body of factual information, providing material which a later age employed to reconstruct the categories of biological thought.

In such a situation, taxonomy inevitably assumed great importance. Whereas Gaspard Bauhin described some six thousand different species in his herbal from the early 17th century, John Ray included over eighteen thousand species in his *Historia plantarum generalis,** which appeared at the end of the century. Some system of classification was essential to organize such a body of data. By 1750, when Linnaeus' work marked a turning point in botany, no less than twenty-five systems had been proposed. Most of them were artificial, as botanists are wont to say, seizing arbitrarily on one characteristic as the criterion of classification instead of utilizing the whole plant and its natural affinities to form what is called the natural system. Whatever defects they embodied, the systems did succeed in organizing the immense number of species into manageable patterns, and they did prepare the way for the greater taxonomists of the 18th century.

Botany reached its highest level in the work of the Frenchman, Joseph Pitton de Tournefort (1656–1708), and the Englishman, John Ray (1627–1705). Tournefort was the first systematically to classify the categories higher than genera, dividing all plants into twenty-two classes, which in turn divide into families within which the genera find their place. Ray established the basic distinction of the monocotyledons and dicotyledons (plants which germinate with a single leaf and those which germinate with two). Tournefort contended that the genus is the most important category of classification and reformed nomenclature to express genera with names of one word. Ray insisted equally on the species as the ultimate unit. In the 18th century, Linnaeus drew on both to develop the binomial system of classification, in which plants are divided into genera and species, the two words in their names fully locating them in the system. The systems of both Tournefort and Ray were far from perfect, and botany recognizes Linnaeus above them both as its great taxonomist. The extent of Linnaeus' debt to their work, however, is witness to the contribution of 17th century naturalists.

In the case of zoology, the multiplicity of life forms combined with the very availability of a seemingly satisfactory system to inhibit similar progress. Success in botany was confined largely to plants with the familiar pattern of roots, stems, and leaves; difficult forms such as alga and moss presented unsolved enigmas and were put to the side as imperfect herbs. In contrast, zoology faced a multiplicity of forms which could not be avoided, such as quadrupeds, birds, reptiles, fish, shell-fish

* *General History of Plants.*

and insects, to which microscopical life was added during the century. By apparent good fortune, however, the ancient world had provided in Aristotle a systematizer who reduced the bewildering variety to order. Undoubtedly the existence of the Aristotelian system helps to explain the fact that the 17th century devoted far less attention to zoological taxonomy than to botanical, and another century was to pass before zoology burst out of Aristotelian classification.

How heavily tradition weighed on zoology may be seen in the massive works of Aldrovandi which appeared between 1599 and 1616—in all ten folio volumes with more than seven thousand pages. Alas, most of the erudition was derivative. Of 294 pages devoted to the horse, three or four concerned themselves with its zoological characteristics while the rest presented a compilation of everything that had even been said about the temperament of horses, their use in war, their sympathies and antipathies, and so on. Aldrovandi followed the Aristotelian classification without question. Even though John Ray tried to reform the classification of sanguineous animals (we would say vertebrates) by using comparative studies of the circulatory and respiratory systems, he ended up with five classes virtually identical to Aristotle's. For all the defects it would later reveal, Aristotle's zoological classification did organize knowledge into coherent patterns—like the botanical systems, which were more original because they inherited less.

Taxonomy provided the broad framework within which biological knowledge was organized. Within the framework, detailed investigation of a wide variety of biological problems was carried on. Studies of individual organs filled in the outlines of human anatomy which Vesalius and his successors established during the 16th century. Anatomy today is full of names which commemorate the labors of 17th century investigators—Glisson's capsule, the Malpighian bodies, Wharton's duct, the aqueduct of Sylvius, Brunner's glands. The fact that few laymen have ever heard of the parts thus named testifies to the depth of the 17th century anatomy. Nor was anatomical research confined to the human body. During the second half of the century, similar detailed studies by Claude Perrault, Edward Tyson, and others were devoted to other species. Marcello Malpighi's *Dissertatio de bombyce** (1669) contained the first successful study of the internal organization of insect life. It is true that comparative anatomy did little more than suggest its own possibility during the 17th century, as the failure of taxonomists seri-

* *Treatise on the Silkworm.*

ously to refine Aristotle's classification demonstrates. A beginning, however hesitant, is still a beginning, and comparative anatomy traces its
history to the age of the scientific revolution.

No single thing contributed more to biological research during the
century than the invention of the microscope, apparently in 1624. What
the telescope was to astronomy the microscope was to biology. If Galileo's discovery of new planets (as he called the satellites of Jupiter)
excited the imagination of Europe, the revelation of the microscope, that
wholly unsuspected levels of life exist, not above us, but about us and
within us, stimulated it more. "I have used the Microscope to examine
bees and all their parts," Francesco Stelluti exclaimed in the first publication of microscopical observations. "I have also figured separately
all members thus discovered by me, to my no less joy than marvel,
since they are unknown to Aristotle and to every other naturalist."
Stelluti got magnification of roughly five diameters. By the end of the
century Anthony van Leeuwenhoek realized magnifications approaching
three hundred diameters and observed forms of life Stelluti had not
dreamed of. (See Fig. 5.1.) Even the cynicism of Jonathan Swift reflects the sensation he caused.

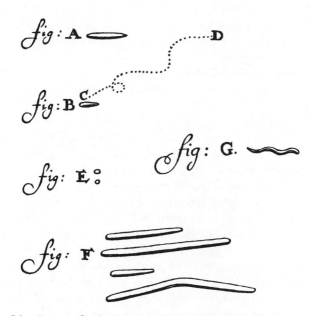

Figure 5.1.   Leeuwenhoek's figures of bacteria from the human mouth.

> "Fleas, so naturalists say,
> Have smaller fleas that on them prey.
> These have smaller still to bite 'em,
> And so proceed ad infinitum."

The second half of the 17th century was the heroic age of microscopy; the early observations were not improved on and seldom equaled before the 1830s. Leeuwenhoek (1632–1723) stands out as a giant even among the heroes. Using single lenses, more magnifying beads than microscopes, he achieved magnifications not repeated for over a century. Swift's fleas on fleas referred to Leeuwenhoek's little animals, infusoria and rotatoria observed in rain water. "When these *animalcula* or living Atoms did move, they put forth two little horns, continually moving themselves. The place between these two horns was flat, though the rest of the body was roundish, sharpning a little towards the end, where they had a tayl, near four times the length of the whole body, of the thickness (by my Microscope) of a Spiders-web; at the end of which appear'd a globul, of the bigness of one of those which made up the body." He observed spermatozoa, and he discovered the corpuscles of the blood,—"flat oval particles, swiming in a clear liquor." Just as more than a century had to pass before the observations were improved upon, so an equal period had to pass before their full significance was realized. Meanwhile they constituted a magnificent addition to the corpus of biological knowledge.

The immense expansion of biological knowledge—an expansion quite unequalled by the expansion of physical knowledge—was accompanied by a reconsideration of the nature of life as the mechanical philosophy extended its influence over the last stronghold of Aristotelianism. A comparison of two contemporaries, William Harvey and René Descartes, both of whom played major roles in the biological thought of the 17th century, reveals something of the complexities of the relationship between biology and the mechanical philosophy.

In an age when English medical education remained primitive, William Harvey (1578–1657) travelled to Padua in 1600 to study for his medical degree. Padua was the foremost center of medical science in Europe. There Vesalius had dissected and lectured, and there the succession of eminent anatomists who followed him was represented by Fabricius of Aquapendente during the period of Harvey's stay. The result of half a century's careful study conspired to raise doubts in Harvey's mind about the function and operation of the heart.

According to prevailing Galenic physiology, the liver is the primary organ of the body. (See Fig. 5.2.) Here food receives its first elaboration, being converted to blood. Imbued with natural spirits, blood flows from the liver through the system of veins to the organs and parts of the body, where it is absorbed as food. Part of the blood enters the right ventricle of the heart and seeps through pores in the septum, the partition separating the two ventricles, to enter the left ventricle where it undergoes a second elaboration in the presence of air, which enters from the lungs. What emerges from the left ventricle to be carried throughout the body by the arterial system is vital spirits, a fluid as different from blood as blood is from food. Part of the vital spirits that ascend to the brain undergo a third elaboration there and are converted to animal spirits, which are distributed through the nerves.

Galenic physiology, thus briefly summarized, held its ascendance partly because it expressed itself in conceptions acceptable to a premechanical age and partly because the functions it assigned to organs conformed to the facts of dissection. Rather, they conformed to the facts until Vesalius tried to find the pores in the septum and failed. Others

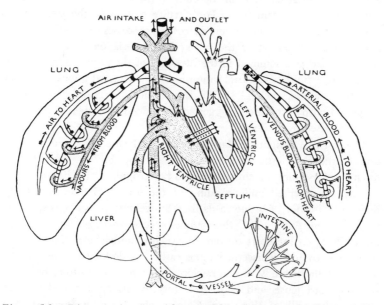

Figure 5.2. Diagram of action of heart and blood vessels according to Galen. Inset on right is a diagram of the circulation in the lung according to Servetus.

after him failed equally to find them. Fortunately, however, a second discovery made it possible to salvage Galenic physiology with minor revisions. Anatomists found that blood passes from the right ventricle to the left through the lungs. Those who established it considered the pulmonary transit as an alternate route now that passage through the septum was acknowledged to be closed. The venous and the arterial systems continued to be separate, each conveying a unique fluid throughout the body. Galenic physiology remained essentially intact.

Nor was it challenged by Fabricius' discovery of membranous pockets in the veins. We call the membranes "valves" and say that they prevent flow toward the extremities. Fabricius called them *"ostiola,"* little doors, and held that they merely obstruct flow in that direction, mitigating its excessive force so that the soft walls of the veins are not ruptured, slowing its rate sufficiently to allow the members to be nourished.

One further influence of Padua, its prevailing Aristotelianism, exerted itself on Harvey. In physiology, Aristotle had asserted the primacy of the heart in contrast to the primacy of the liver in Galenic physiology. Early in the 17th century, there was a great deal of talk among Aristotelians likening the heart in the body to the sun in the cosmos. Life-giving heat flows from both. The circular motion of the sun around the earth plays a significant role in cosmic processes. Should there not be a similar circulation of the heart? The association of a circulation with the heart was common in the literature of the period, although the word "circulation" held various meanings. One equated it with a cyclical repeating motion, such as systole and diastole. A chemical meaning, connected with distillation, suggested that blood is heated in the heart and condensed in the lungs.

The essential insight of Harvey was to apply the concept of circulation to the now established facts of anatomy and to insist that a mechanical meaning of circulation also be recognized. He began by reversing the accepted understanding of the heart's motion. Observing dogs in vivisection, (as one reads the 17th century physiologists, one is sometimes surprised that the canine species managed to survive) especially when the heart slowed down with approaching death so that its motion could be discerned more easily, he decided that the active motion of the heart is its contraction, the systole. In systole, he could feel the heart tense, and as it drew together its apex was thrust out striking the wall of the ·chest. Galenic physiology, in contrast, had considered expansion, diastole, to be the motion of the heart. When it expanded, the heart attracted, or "drew" a quantity of blood into it.

The attraction was understood, not in mechanical terms analogous to a vacuum pump, but in terms reminiscent of the sympathies of Renaissance Naturalism. This conception, Harvey insisted, was wrong. The "intrinsic motion of the heart is not the diastole but the systole."

The further question immediately arose: what happens to the blood in the heart? Valves at the entrance to each ventricle are arranged in such a way that the blood cannot flow out through the passage by which it enters; valves at the exits prevent its re-entering again once it has left. (See Fig. 5.3.) Over and over, the same action repeats, each stroke thrusting a new quantity of blood after the one before. Blood from the right ventricle, of course, is driven through the lungs and into the left. What happens to that forced out of the left? To his insistence on the mechanical necessities of the heart, Harvey now added another argument wholly typical of 17th century science. By measuring the capacity of a dissected heart, he determined that a ventricle holds more than two ounces of blood; to be on the safe side, he assumed a maximum capacity of two ounces. Suppose that a fourth of it is driven out by each contraction; to be on the safe side, he set it as low as an eighth. And suppose the heart beats a thousand times in half an hour —again a figure deliberately too low. According to our present information, Harvey's calculation of the blood discharged by the heart was less than three percent of the true quantity. Never mind; his purpose was not measurement as such, but the polemic value of a quantitative argument deliberately understated. By a simple calculation, he showed that even with the underestimates the heart discharges more blood into the arteries in half an hour than the entire body contains. Where can it possibly go, but back to the heart by another route?

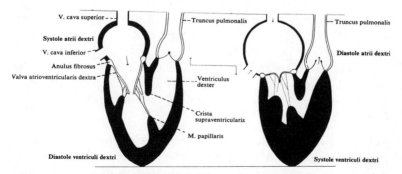

Figure 5.3. A modern diagram shows the action of the valves of the right ventricle in diastole and systole.

Harvey had demonstrated the necessity of circulation. The problem was to demonstrate as well that circulation is a fact. Without a microscope, he was unable to observe the capillaries which connect the arterial system to the venous. Nevertheless, by an ingenious experiment on himself, Harvey was able to show that the blood does pass from the arteries to the veins. Applying what was called a perfect ligature to his arm, he cut off both the veins and the artery. The arm gradually grew cold but did not change color; above the ligature the artery filled and throbbed. Loosening the ligature enough to free the artery while the veins remained blocked, he felt the surge of warmth as fresh blood was forced through his arm. Immediately, the arm became purple and the veins swelled visibly below the ligature. They had not been filled from the venous system which remained cut off; the blood had to reach them from the arteries.

The essence of Harvey's demonstration of the circulation of the blood lay in his attention to the mechanical necessities of the vascular system. On this question, the mechanical mode of thought, so spontaneous to the 17th century mind, could offer assistance to biological science. The heart functions as a pump moving a fluid through a closed circuit of conduits, a system reminiscent of the waterworks which ran the elaborate fountains admired by 17th century monarchs. As a paragraph among his lecture notes says,

"From the structure of the heart it is clear that the blood is constantly carried through the lungs into the aorta *as by two clacks* [valves] *of a water bellows to rayse water.*"

The same William Harvey in the same book which expounded the circulation of the blood also called the heart "the beginning of life."

"The heart is the sun of the microcosm, even as the sun in his turn might well be designated the heart of the world; for it is the heart by whose virtue and pulse the blood is moved, perfected, and made apt to nourish, and is preserved from corruption and coagulation; it is the household divinity which, discharging its function, nourishes, cherishes, quickens the whole body, and is indeed the foundation of life, the source of all action."

Although Harvey saw the heart as a pump, he did not see it solely as a pump, or even primarily as a pump. The circulation of the blood, the mechanical effect of a machine neatly contrived, exists to serve an end which is not mechanical. Its circulation recalls the cycle of evaporation and rain which emulates, as it is caused by, the "circular motion of the superior bodies" by which the generation of all living things is produced.

"And so, in all likelihood, does it come to pass in the body, through the motion of the blood, that the various parts are nourished, cherished, quickened by the warmer, more perfect, vaporous, spirituous, and, as I may say, alimentive blood; which, on the contrary, in contact with these parts, becomes cooled, coagulated, and, so to speak, effete; whence it returns to its sovereign the heart, as if to its source, or to the inmost home of the body, there to recover its state of excellence or perfection. Here it resumes its due fluidity, and receives an infusion of natural heat—powerful, fervid, a kind of treasury of life, and is impregnated with spirits, it might be said with balsam, and thence it is again dispersed."

Harvey was a thorough-going Aristotelian who saw in the circulation of the blood one aspect of the primacy of the heart. Unlike his master, he insisted on the role of the blood as well, heart and blood together forming a single functioning unit which is the very seat of life, a basis which has nothing whatever to do with mechanisms and matter. The blood is a spiritual substance.

"It is also celestial, for nature, the soul, that which answers to the essence of the stars, is the inmate of the spirit, in other words, it is something analogous to heaven, the instrument of heaven, vicarious of heaven."

In his study of the generation of animals, Harvey had observed a pulsing point of blood as the first sign of life in the embryo. In death, a palpitation of the blood was the last living act—"nature in death, retracing her steps, reverts to whence she had set out, returns at the end of her course to the goal whence she had started."

For Harvey as for Aristotle, then, circulation had manifold meaning, reproducing the cyclical regeneration which is the means to the preservation of the cosmos and all it contains. In the cyclic alteration of birth, reproduction, and death, he saw another reflection and embodiment of the eternal orbits which determine the generation and corruption of terrestrial beings. By describing the circuit, the species achieves immortality;

"now pullet, now egg, the series is continued in perpetuity; from frail and perishing individuals an immortal species is engendered. By these, and means like to these, do we see many inferior or terrestrial things brought to emulate the perpetuity of superior or celestial things. And whether we say, or do not say, that the vital principle inheres in the egg, it still plainly appears, from the circuit indicated, that there must be some principle influencing this revolution from the fowl to the egg and from the egg back to the fowl, which gives them perpetuity."

So also some principle must govern the circulation of the blood. The mechanical necessity of circulation expresses only its material conditions. But blood is a spiritual fluid, the bearer of the vital principle on which life depends. Its true circulation is the cycle of renewal and decline. It leaves the heart warm and vital, bearing life to the extremities, and returns coagulated and effete to be restored. In its circulation, the blood repeats in the microcosm the cosmic cycle of generation and corruption, and in its repetition preserves the life of the individual.

When Harvey's *De motu cordis et sanguinis** was published in 1628, Descartes was already at work on the reconstruction of natural philosophy. Inevitably, Harvey's discovery interested him; inevitably, he comprehended it in his own terms. The notion that blood moves in a closed circuit, an idea that corresponded to his exposition of motion in a plenum, did not fail to catch his eye. Consequently, when his *Discourse on Method* appeared ten years after Harvey's book, it included an exposition of the circulation of the blood as an example of a purely mechanical physiological process.

"And that there may be less difficulty in understanding what I am about to say on this subject," he counselled as he began, "I advise those who are not versed in Anatomy, before they commence the perusal of these observations, to take the trouble of getting dissected in their presence the heart of some large animal possessed of lungs, (for this is throughout sufficiently like the human)." The advice probably fell as incongruously on the ears of the 17th century reader as it does on those of the 20th century one. For the benefit of those readers who had no one to cut open a heart for them and preferred not to do it themselves, Descartes described its structure, laying stress on the valves which "readily permit the blood to pass, but preclude its return." He remarked as well that the heart has more heat than the rest of the body. In it there is kindled what he called "one of those fires without light, not different from the heat in hay that has been heaped together before it is

---

* *On the Motion of the Heart and Blood.*

dry, or that which causes fermentation in new wines." He understood such a fermentation as a mechanical process, of course.

When portions of blood enter the two ventricles, they "are immediately rarefied, and dilated by the heat they meet with."

"In this way they cause the whole heart to expand, and at the same time press home and shut the five small valves that are at the entrances of the two vessels from which they flow, and thus prevent any more blood from coming down into the heart, and becoming more and more rarefied, they push open the six small valves that are in the orifices of the other two vessels, through which they pass out, causing in this way all the branches of the arterial vein and of the grand artery to expand almost simultaneously with the heart—which immediately thereafter begins to contract, as do also the arteries, because the blood that has entered them has cooled, and the six small valves close, and the five of the hollow vein and of the venous artery open anew and allow a passage to other two drops of blood, which cause the heart and the arteries again to expand as before."

Those who do not appreciate the force of mathematical demonstrations, he added, must be warned "that the motion which I have now explained follows as necessarily from the very arrangement of the parts, which may be observed in the heart by the eye alone, and from the heat which may be felt with the fingers, and from the nature of the blood as learned from experience, as does the motion of a clock from the power, the situation, and shape of its counterweights and wheels."

What Descartes had done was to appropriate Harvey's discovery but systematically to eliminate Harvey's vitalism which he regarded as occult. In his *Traité de l'homme** he described a machine that performs all the physiological functions of man—circulation, digestion, nourishment and growth, perception.

"I want you to consider [he concluded] that all these functions in this machine follow naturally from the disposition of its organs alone, just as the movements of a clock or another automat follow from the disposition of its counterweights and wheels; so that to explain its functions it is not necessary to imagine a vegetative or sensitive soul in the machine, or any other principle of movement and life other than its blood and spirits agitated by the fire which burns continually in its heart and which differs in nothing from all the fires in inanimate bodies."

* *Treatise on Man.*

It is not necessary to imagine a principle of life—here was the crux of Cartesian physiology. Life itself was an alien presence in a mechanical world. Indeed, it was not a presence at all, but a mere appearance to be explained away with other occult properties.

To say that Descartes appropriated Harvey's discovery is only half true until we add that he bowdlerized it egregiously in the process. Determined to eliminate any mysterious entity such as life, he insisted on deriving the motion of the heart from known physical processes; in doing so, he turned the heart into a teakettle. More than that, the physiology of the radical innovator represented a reactionary step backward in comparison to that of Harvey, the conservative Aristotelian. Whereas Harvey established the fundamental role of the systole, Descartes' vaporization returned to the Galenic diastole. He accepted circulation, it is true, but the vaporized blood which leaves the heart in his system recalls Galen's vital spirits, and he described a separation in the brain of the most subtle particles of the blood to form the animal spirits which circulate through the nerves. Cartesian physiology was basically Galenic physiology reattired in the robes of mechanical philosophy. A lifetime's contemplation of vital phenomena left Harvey convinced that they could not be reduced to material explanations. For a priori reasons that did not derive in any way from biological considerations, Descartes vulgarized Harvey's work in order more easily to mechanize it. In the process, he even lost the principal elements of Harvey's mechanical treatment of cardiac motion. It was not a happy augury for the contribution of mechanical philosophy to biological science.

Nevertheless, Descartes determined the tone of biological studies in the later 17th century far more than Harvey did, and there developed a school of mechanical biology known as iatromechanics. Biology remained more richly varied than chemistry, and iatromechanics never dominated it to the extent that mechanism came to dominate chemistry. Iatromechanism was more than a factor in the biological science of the late 17th century, however; it was the distinctive feature.

De motu animalium (1680–1) by Giovanni Alfonso Borelli (1608–1679) ranks among the best products of iatromechanics. First for man, and then for other animals including birds and fish, Borelli applied the principles of simple machines to the analysis of various movements. (See Fig. 5.4.) Consider, for example, a man crouched and ready to spring into the air. Borelli examined the position of the muscles that must contract and their connection to the skeleton. His basic in-

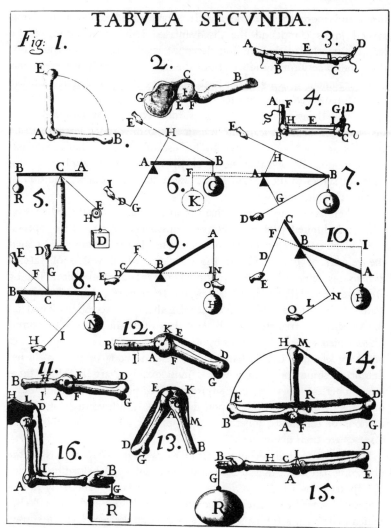

Figure 5.4. A set of diagrams from Borelli's work illustrates the mechanical principles that he applied to the operation of muscles and joints.

sight, both here and in other cases, was that the muscles work always at a considerable mechanical disadvantage. Treating the bone as a lever with the joint as its fulcrum, he showed that the muscle which supplies the motive force connects very close to the fulcrum, whereas the load is generally placed near the other end of a bone with a lever arm ten times and more that of the muscle. Complicated motions involving several joints compound the disadvantage. Thus in the case of the leap, he concluded that the muscles must exert a force over four hundred and twenty times the weight of the man just to pull him erect, and by an argument which will not bear close scrutiny, he concluded further that a force seven times greater is required to project the man into the air. In all, then, a man must exert a force twenty-nine hundred times his weight in order to leap into the air. As in the case of leaping, all of Borelli's analyses were vitiated by his use of static equilibrium to examine motion. Beyond that, however, his willingness to apply the principles of statics to the human frame was a sound, if minor, addition to biological understanding.

Neither Borelli nor iatromechanists in general were satisfied to stop with such limited problems. Harvey's discovery of circulation opened a broad field for mechanical investigations. Iatromechanists calculated the velocity of the blood and the resistance which vessels of various dimensions offer to it. They proposed to explain animal heat, not by Descartes' fire without flame, but by the friction of the blood with the walls of the arteries. They constructed a theory of secretions based on the velocity of circulating fluids, and they filled the body with porous filters which separated particles by sizes and shapes. It is generally recognized, proclaimed Dr. Richard Mead, that the body of man is "a hydraulic machine contrived with the most exquisite art, in which there are numberless tubes properly adjusted and disposed for the conveyance of fluids of different kinds. Upon the whole, health consists of regular motions of the fluids, together with a proper state of the solids, and diseases are their aberrations."

Such a view of life could not fail to color the observations of the observing naturalist. In at least two areas of biology, it helped to obstruct the appreciation of discoveries of major importance. The early microscopists observed the cellular structure of wood. Our very word "cell," which plays such a fundamental role in biological science, was first used as a biological term by Robert Hooke (1635–1703) in *Micrographia*, (1665). Observing a piece of cork under a microscope, he was reminded of a honeycomb and referred to what he saw as pores or cells. The word "pores" was more expressive of Hooke's interpretation. They seem, he said, "to be the channels or pipes through which the *Succus nutritius,* or natural juices of Vegetables are convey'd and seem to correspond to the veins, arteries, and other Vessels in sensible creatures." He even looked for valves to control the direction of flow, and though he failed to observe them, he thought it probable nevertheless that nature had not failed to provide such "appropriated Instruments and contrivances" to achieve her purposes.

The whole tenor of 17th century thought inclined the microscopists to see in this discovery, not the ultimate unit of life, but pipes suitable to carry fluids. As Nehemiah Grew, who extended Hooke's initial observations into a complete theory of vegetable physiology, asked: "to what end are Vessels, but for the conveyance of Liquor?" Additional irony derives from the fact that microscopists also observed unicellular creatures such as spermatozoa. They could not even dream that their "little animals" bore any relation to the pores observed in plants.

A much more complicated story revolves about the study of embryology. From the ancient world the 17th century did not inherit a unified theory of generation, but rather different theories for different classes of beings. The generation of viviparous quadrupeds (and man) obviously differed from that of oviparous animals. Insects were held to generate spontaneously from decaying material, and the reproduction of plants was another matter altogether. It was the work of William Harvey, one of the first great embryologists of the modern world, as well as the discoverer of circulation, which attempted to comprehend the generation of all animals in common terms. The frontispiece of his book, *De generatione animalium** (1651), shows Zeus opening an egg from which animals of all sorts, including a human, emerge, and on the egg appears the legend "Ex ovo omnia" or, as he stated the same idea in the treatise, "An egg is the common origin of all animals." On close examination, the word "egg" turns out to be highly ambiguous. In the case of oviparous animals, it is definite enough. Harvey never comprehended the function of the organs that we call ovaries in viviparous animals, however. What he called the egg of the deer was the amniotic sac, in which an embryo had been developing for several weeks. In the case of insects, it was the cocoon from which the butterfly emerges. By egg then, he meant, not a product of a female ovary, but what he also called a "primordium," a first matter or first beginning however produced. It was a broad enough concept to embrace even the spontaneous generation of insects, which Harvey did not question.

Nevertheless, Harvey's formula embodied a considerable generalization. Whatever the ambiguity in his meaning of egg, he had attempted to comprehend all generation under one common pattern. Even the seed

* *On the Generation of Animals.*

of a plant could be considered a *primordium*. The details of generation may vary from species to species, but in all of them the egg represents one point in the eternal cycle of reproduction by which the species is preserved.

An egg, the origin of every being, was to Harvey an homogeneous point of matter which an indwelling formative principle molds and converts into an articulated individual able to produce, as its ultimate act, an homogeneous point of matter, the primordium of another generation. In his examination of a doe, Harvey could find no trace of male semen in the uterus, and the egg of the deer was first visible to him seven weeks after coition. Obviously, the male semen cannot play a material part in generation. Harvey described its action by the word "contagion," an immaterial influence which lingers and stimulates the dormant egg. Once stimulated and awakened to activity, the egg had within it both an indwelling principle and material for it to work on. Harvey coined the word "epigenesis" to describe the process he observed in the generation of chickens. In an egg opened three days after it was laid, he saw a pulsing point of blood which became the heart, the first organ to be formed and the center from which the rest of the chicken was generated. Epigenesis was the natural expression of Harvey's vitalism, a creative generation under the guidance of the formative virtue which embodies the divine idea of the species.

Descartes was ready to mechanize epigenesis along with the rest of life. In *La description du corps humain,** he described how male and female semen ferment when joined, and how the resulting motions, by mechanical necessity, build the heart, the circulatory system, and so on. The 17th century considered the account to be arrant nonsense, just as we do, and an alternative embryology suggested by Gassendi won a wider audience. To Gassendi, the fundamental act of generation was the production of a seed. Both in plants and in animals, the seed is a tiny body containing particles from all parts of the individual. He spoke sometimes of a soul in the seed, but since the soul was itself composed of aethereal matter, it did not dilute the essential mechanism of the account. The controlling factor in generation is the attraction of like for like, an idea uncomfortably reminiscent of Renaissance Naturalism but seemingly capable of translation into harmonious shapes and motions. In a seed, like particles (deriving from the same parts) come together, and they attract other like particles from the

---

* *The Description of the Human Body.*

food available. Hence, in some sense, the product of generation is already present in the seed. As Gassendi declared, "the seed contains the thing itself, but contains it as rudiments not yet unfolded."

The term "preformation" is attached to this conception of generation. Epigenesis considered generation as a creative process in which the formative virtue molds and alters the material present to it, evoking heterogeneity from homogeneity. Preformation, on the other hand, asserted that heterogeneity must be present from the beginning and that generation is merely the process of its evolution (literally, unfolding) or development (literally, emerging from envelopes, or uncovering). "Heterogeneity" was a term readily understood by atomists, who likewise believed that it is present from the beginning in the form of particles of different shapes. Not merely in embryology but in general the mechanical philosophy regarded the formation of all individual things as a process by which suitable pre-existing particles are fitted together. Descartes' attempt to mechanize epigenesis had been an obvious failure, but preformation offered a mechanical alternative to the unacceptable idea of a formative virtue.

Marcello Malpighi (1628–1694), perhaps the greatest embryologist of the century, elaborated Gassendi's account. By perfecting a technique of removing the cicatrix from a freshly opened egg and spreading it on glass, Malpighi was able to introduce the microscope into embryology. Just six hours after the egg was laid, he discerned the cephalic region and the spine. Outlines of vertebrae appeared after twelve hours. On the second day, he saw the beating heart, which Harvey, without a microscope, had seen only on the fourth. With the heart, he saw the head and the beginnings of eyes. Naturalists, he declared, have sought to discover the genesis of separate parts in different stages; "while we are studying attentively the genesis of animals from the egg, lo! in the egg itself we behold the animal already almost formed."

When Malpighi came to study the generation of chickens, he was already an experienced investigator both of plants and of silkworms. In the silkworm, he had found the wings and the antennae of the butterfly already existing as rudiments in the body of the caterpillar, and in a bud, he had discovered "a compendium of the not yet unfolded plant." His mind was thus prepared to find the chicken present in the egg from the beginning. It was present, however, as rudiments. He spoke of sacules and vesicles within which different parts develop. Walled off from the rest of the egg by membranes which acts as sieves, the vesicles "admit appropriate matter, which is consumed in the construction of the parts," and when the vesicles are joined together, the structure of the animal appears. Clearly, the filtering action of the porous membrane was a rendition of Gassendi's attraction of like for like, just as his term "rudiments" repeated Gassendi's phrase.

Whereas Malpighi was primarily a skillful observer, others were more concerned with systematizing, and in them the subtlety of Malpighi's preformationism was cast to the winds. Eggs, Swammerdam pronounced, are not transformed into chickens, "but grow to be such by the expansion of parts already formed." "There is never any generation in nature," he added, "but only a stretching or a growth of parts." If there is never any generation in nature, then eggs themselves cannot be generated. In the chicken, preformed in the egg, there are preformed eggs as well, and of course in those eggs preformed chickens with their preformed eggs.

"In eggs, so naturalists say . . ."

At the end of the 17th century, embryology produced the theory of *emboîtement* which held, for example, that the entire human race was present already in Eve.

That the theory of *emboîtement* included Eve and the human race as well as chickens was due to further discoveries which seemed at the time to confirm preformationism. In 1667, Nicholas Steno discovered the ovaries, filled with eggs, in the dog fish, a viviparous creature. Five years later, Regnier de Graaf (1641–1673) discovered vesicles on the female testicles (as they were then called) of rabbits, dogs, cows, and humans. He took the vesicles to be eggs and asserted that the so-called testicles are in fact ovaries. In a brilliant set of experiments with pregnant rabbits, he found a constant numerical identity between the number of embryos in the uterus and the number of yellow bodies on the ovaries—the *corpora lutea* left by the vesicles after ovulation. Although de Graaf mistook the vesicle for the egg, (the mammalian egg is so small that it was not observed until the 19th century), his interpretation of what he discovered was essentially correct, and we continue to commemorate it with the name Graafian follicle. Harvey's dictum now acquired a new and more exact meaning; viviparous mammals are indeed born of eggs. Preformationism had established itself on the study of generation in eggs; ovism, as the doctrine of the universality of generation from eggs was called, appeared to lend it powerful support.

The uncontested reign of ovism lasted exactly five years. What the microscope gave the microscope took away. In 1677, Leeuwenhoek observed spermatozoa. (See Fig. 5.5.)

"These *animalicula* were smaller than the corpuscles that make the blood red, so that I estimate a million of them are not equal in size to a large grain of sand. They had roundish bodies, blunt in front, but ending in a point at the rear; they were endowed with a thin transparent tail five or six times as long as the body and about one twenty-fifth as thick, so that I can best compare their shape to a small radish with a long root. They moved forward with a serpentine motion of the tail, like eels swimming in water."

Ovism, it now appeared, was a monstrous mistake. The passive egg could be nothing but food for the true agents of reproduction, the manifestly vital animacules or, as he also called them, the "spermatic worms" of the male semen. A Swedish doctor found the new doctrine

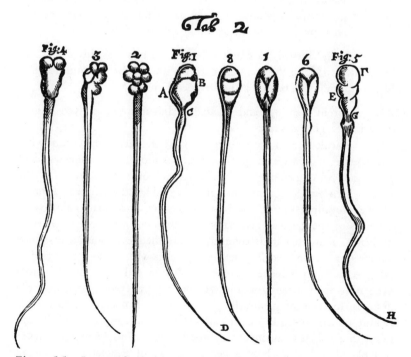

Figure 5.5. Leeuwenhoek's drawing of spermatozoa. 1–4 represent human spermatozoa, 5–8 canine spermatozoa.

more conformable to the dignity of man. Niklaas Hartsoeker (1656–1725) showed the absurdity of ovism by calculating that an original egg would be larger than one destined for fertilization sixty centuries later (since the creation of the world was popularly placed in 4004 B.C., he was comparing Eve with his own generation) by a factor of $10^{30,000}$.

One might suspect then that animaculism (as the new doctrine was called) rejected preformation. Nothing could be further from the truth. The same factors that made preformation attractive to the ovists made it attractive to the animaculists. The same Hartsoeker who demonstrated the absurdity of ovism failed to see that animaculism suffered from the identical problem.

"It can be said that each animal, actually and in miniature, contains and shields in a delicate and tender membrane a male or female animal of the same species, as that in the semen of which it is found."

He even published a picture of an homunculus all curled up in the head of a spermatozoon. (See Fig. 5.6.) As a satirical reply, a French doctor, François de Plantade, published a similar figure and told how he had observed an homunculus in the act of sloughing off its envelope.

"He clearly showed, bare and exposed, his two legs, his thighs, his belly, his two arms; the membrane drawn toward the top coiffured him like a capuchin. He paused as he stripped himself."

Alas, the irony was lost, and Plantade's drawing was received as confirmation of Hartsoeker's.

It is difficult to read the embryologists of the late 17th century without a sense of bewilderment. Their contribution to the knowledge of generation was immense. To the discovery of spermatozoa and the virtual discovery of mammalian eggs, they added the effective disproof of the prevailing notion that worms, insects, and small animals are products of spontaneous generation and the demonstration of sexuality in plants. Francesco Redi conducted controlled experiments in which worms appeared in decaying meat open to flies whereas none appeared in samples carefully screened. He concluded that the worms, far from generating spontaneously, are larvae which grow from eggs laid in the flesh. In the case of plants, R. J. Camerarius demonstrated that seeds require pollen from the stamens in order to reach maturity. He recognized that pollen is analogous to male semen. Thus biological science was placed within reach of a general theory of generation embracing all living forms. What was actually produced in preformationism, how-

Figure 5.6. Hartsoeker's conception of how an homunculus ought to look in an animacula in the sperm.

ever, was a theory unable to account satisfactorily for the most obvious fact of generation, that offspring can and do inherit characteristics from both parents.

It is tempting to conclude that the mechanical philosophy, with its inability to recognize in generation anything but an unfolding of pre-existent parts, stood between 17th century embryology and the comprehension of its own discoveries. Before we accept such a conclusion, we should recall that Harvey the vitalist, the exponent of epigenesis, was also an ovist who denied any contribution of the male semen to the embryo. For reasons almost diametrically opposed to those of mechanical philosophers, that is, to assert the nonmateriality of generation, Harvey rejected the possibility of material contact between the semen and the egg. More than the mechanical philosophy obstructed comprehension of the new discoveries. A vast range of additional knowledge and understanding of vital processes, not available until the 19th century, was needed before the full import of 17th century discoveries could be realized. We need to remember as well that the discoveries in embryology, as well as many others in the whole field of biology, were in fact made during an age when the mechanical philosophy held sway over scientific thought. However inappropriate its categories for biological understanding, it did not prevent the great expansion of biological knowledge.

Another temptation must equally be resisted—the temptation to greet the iatromechanists as early biophysicists and biochemists. Iatromechanism did not arise from the demands of biological study; it was far more the puppet regime set up by the mechanical philosophy's invasion. In isolated problems—the circulation of the blood is the classic example —mechanical modes of thought, the ability to see the mechanical necessity in a vital process, could lead to new insights. Harvey himself, however, was a vitalist, not a mechanist. For the most part, iatromechanics was simply irrelevant to biology. It did not prevent the vital work of detailed observations; it contributed almost nothing toward understanding what was seen. Beside the subtlety of biological processes, the 17th century's mechanical philosophy was crudity itself. Above all, it lacked a sophisticated chemistry which has turned out to be a prerequisite for the rapprochement of the physical and biological sciences. One can only wonder in amazement that the mechanical explanations were considered adequate to the biological facts, and in fact iatromechanics made no significant discovery whatever.

CHAPTER VI

Organization of the Scientific Enterprise

MORE THAN SIMPLY a reformulation of scientific conceptions occurred in the 17th century, even though the reformulation of
conceptions was radical enough to warrant the name "revolution" that
is frequently applied to it. Science as an organized social activity also
appeared. Obviously, earlier periods had witnessed a great deal of
scientific activity. It is difficult to distinguish science from philosophy
before the 17th century, however, and it is equally difficult to describe
many men primarily as scientists. The existence of a Leibniz indicates
that the compartmentalization of what we now call science was far from
complete at the end of the 17th century. Nevertheless, by that time,
Western Europe contained, not just a few, but whole groups of men
whom we label without hesitation as scientists. Moreover, they were
not working in isolation as individuals, but had organized societies which
placed them in effective communication with large numbers of men
engaged in the same pursuit. On the ground once trod by prophets an
organized church now stood.

The 20th century learns with surprise that the word "university" did
not appear in the title of that church. We are accustomed to think of
universities as the principal centers, or at least as being among the
principal centers, of scientific research. A similar situation had existed
in the Middle Ages, when virtually all intellectual activity, including
science, had been located within university walls. A radically different
situation obtained during the 17th century. Not only were the universities of Europe not the foci of scientific activity, not only did natural
science have to develop its own centers of activity independent of the
universities, but the universities were the principal centers of opposition
to the new conception of nature which modern science constructed.

To understand their relation to modern science, one must remember the circumstances that had called the European universities into being and the function they existed to perform. The acquisition of the corpus of Aristotelian philosophy in the 13th century had effectively created the university as a center of learning. From the beginning, the institution had been devoted to the explication and amplification of Aristotle, and the academic circles of Europe had built a vested interest in the maintenance of his philosophy. From the beginning, the university had also been connected with the Catholic Church. When the Church was the primary receptacle of learning, the university could hardly exist in independence of it. The Church did not impose its will on an institution existing outside it; quite the contrary, the Church created and fostered the university as the foremost institution of learning in a society which would otherwise have had nothing like it. All of the teaching masters in the European universities were in orders, and most of the students were preparing for ecclesiastical careers. In the medieval universities, Aristotle was baptized and christianized, and they attached to him the label, "the philosopher," by which he was known in countless treatises. By the year 1600, very few of the essential features of the institutions had changed. The influence of the Renaissance had apparently introduced other classical authors into the curriculum, but the universities were not the centers of humanistic learning. In Protestant areas, the universities came to serve other denominations without further significant change. As the sons of the nobility aspired to polite learning, the exclusively clerical nature of the institutions began to wane, but their ecclesiastical function did not cease by any means. Thus in 1600, the universities gathered within their walls a group of highly trained intellectuals who were less apt to welcome the appearance of modern science than to regard it as a threat both to sound philosophy and to inspired religion.

Galileo may serve as an example of the relation of science to the university. He began his professional career as the professor of mathematics at the University of Pisa, and in 1592 he moved on to a similar chair in the most important Italian university, at Padua. Throughout the 16th century, Padua had been a center of scientific learning, the home of a series of philosophers whose writings on logic contributed significantly to the philosophical foundations of the scientific method. Their work was firmly based on Aristotle's logic and occasioned no challenge to the prevailing tradition. In contrast, Galileo concerned himself, not with logic, but with cosmology and mechanics, and as we have seen, his work burst the framework of Aristotelian science. It is true that Galileo occupied the chair at Padua for eighteen years, the most creative period of his life, during which he built the structure of his mechanics, and with his telescope helped to destroy the structure of the Aristotelian heavens. In the end, however, he left Padua for Florence and published his great works, the *Dialogue* and the *Discourses,* not as a university professor, but as the mathematician of the Grand Duke of Tuscany. The act was symbolic for the 17th century. With the exception of some doctors, virtually none of the leading scientists held university chairs, and the scientific revolution was created more despite the universities than because of them. If it is symbolic that Galileo left Padua, it is also symbolic that the principal driving force behind the trial which he suffered in Rome came, not from the theologians of the church, but from the entrenched academics who saw in his virulent anti-Aristotelianism a threat to their vested interest in "the philosopher."

If most of the leading scientists worked outside the universities, not all of them did. The greatest of them all, Isaac Newton, occupied the Lucasian chair of mathematics at Cambridge throughout the creative period of his career. While he was the Lucasian professor (and during the five years before his appointment, when he was also at the university), Newton discovered the calculus, the composition of white light, and the law of universal gravitation. Nevertheless, Newton's case does not contradict the assertion that the universities were not centers of scientific activity in the 17th century. It is true that Newton did not face the hostility Galileo had faced; Cambridge at the end of the century was not Padua at its beginning. As a scientist, however, he played no essential role in the educational life of the university. He expounded his discoveries in both optics and mechanics from the podium before they were published. Not one shred of evidence suggests that the lectures were comprehended or that they aroused any response, and a number of indications suggest that they may frequently have gone literally unheard. How indeed could it have been otherwise? Nothing in the standard curriculum prepared the undergraduates for his lectures; the tutorial system in the colleges, the backbone of the instructional method, was directed toward wholly different ends. A Nobel laureate lecturing on his research to entering freshmen in an American university today would be less incongruous than Newton announcing his discoveries to the Cambridge undergraduates of the 17th century. Although he held the university's respect and emerged as a leader of the resistance to the king's effort to subvert Cambridge in the period before the

Glorious Revolution, as a scientist Newton existed in virtual isolation at Cambridge. Contact with the Royal Society in London led to the publication of his work; nothing in Cambridge acted as a similar stimulant.

The English universities, far from being retrogressive in comparison to others in their reception of the new science, were as advanced as any in Europe. In Cambridge, the Lucasian chair of mathematics was established in 1663. Oxford had preceded Cambridge by nearly half a century. Sir Henry Saville had endowed professorships (naturally Savilian professorships) in geometry and astronomy in 1619, and his son-in-law had followed with a Sedleian chair of natural philosophy two years later. On the whole, able men held the positions during the century. We have already seen that Newton did not make Cambridge a scientific center, and the same can be said for the Savilian professors at Oxford. Meanwhile, denunciations of the universities, stressing the continued domination of traditional learning which seemed empty and pointless to those who pressed the charges, filled the air from one end of the century to the other. It is worth remembering that nearly every scientist of importance during the century was a university product. It remains true, nevertheless, that science did not seriously penetrate either the common rooms of the colleges or the curricula of the universities. When the century closed, the traditional curriculum dating from the Middle Ages had not been systematically replaced.

What was true of the English universities was at least as true of the continental ones and in most cases more so. With the principal seats of learning effectively barred to it, the scientific movement created its own institutions, not educational institutions, but organizations which made science a sociological phenomenon as well as an intellectual one. The 17th century witnessed the birth of the scientific societies.

The earliest known organization that might be called a scientific society was the *Accademia dei Lincei* (The Academy of the Lynx) which flourished in Rome in the early part of the century. Galileo was a member of it, and when in his *Dialogue* he had his mouthpiece Salviati mention the "Academician," he was identifying himself by his membership in the Academy of the Lynx. Informal in structure and patterned after literary groups among the Italian humanists, the Academy of the Lynx was a gathering of like-minded friends at which matters of natural philosophy could be discussed. After it ceased to exist about 1630, another similar group was organized in Florence in the middle of the century under the sponsorship of the Medicean duke. The *Accademia* *del Cimento* (Academy of Experiment) was devoted, as its name implies, to exact experimental investigation of the questions exercising natural philosophy at the time. More structured than the Academy of the Lynx, it engaged primarily in corporate experimentation, which it published in a volume entitled *Saggi di naturali esperienze** (1667).

In the other countries of western Europe, informal groups similar to the Academy of the Lynx appeared during the first half of the century. Father Mersenne, a Minim Friar, made himself the center of the considerable circle of Frenchmen who set the pace of European science for a brief period in the middle of the century. An heroic correspondent, Mersenne became the communication center, not only of French science, but of European science. Through him, Galileo's work was introduced into northern Europe. Indeed, he was responsible for the first publication of the *Discourses* in the Netherlands when Galileo, confined to house arrest by the Inquisition, did not dare to publish it himself. A few years later, he spread the news of the Torricellian experiment with the vacuum. He encouraged Pascal's experimentation and fostered the publication of his mathematical work. Descartes found in Mersenne his principal avenue of communication with the rest of the learned world. When he composed his metaphysical treatise, *Meditations on First Philosophy,* Mersenne circulated copies of it among the leading philosophers of the day, so that it appeared in its first publication with seven sets of objections and Descartes' replies—an amplification several times larger than the original work. It is hardly too much to say that Mersenne was a scientific society by himself.

After the establishment of the *Académie française* by Richelieu in 1635 provided French literature with an instrument of organization and the purity of the French tongue with a shield of defense, French science began to feel the need for more formal organization. In Habert de Montmor it found a wealthy patron in whose home the Montmor Academy met. Pierre Gassendi presided at the meetings until his death. Here French science came to focus during the 1650s.

One of the meetings of the Montmor Academy is instructive of the function of the early informal groups from which the scientific societies ultimately grew. In 1658, a paper by Christiaan Huygens, then a young man on the threshold of his career, was read to the academy, a paper which unravelled the shape of Saturn by proposing that rings encompass it. The meeting was something in the nature of an event. Several officials

* *Essays of Natural Experiments.*

of the government were present along with abbots of noble birth and doctors of the Sorbonne, and mere scientists were apparently fortunate to find seats in the back row. Which is to say that the early informal societies were engaged as much in a propagandistic effort as in the promotion of research. A new conception of nature and of man's place in it was in process of creation. It challenged common sense and the sophisticated formalization of common sense in the Aristotelian philosophy, the philosophy known to every educated man and received by most. Perhaps the most crucial function of the early societies was to present the new conception of physical nature to the educated public as a viable alternative. The function was imperishably enshrined in the discussion of the three men brought together in Galileo's *Dialogue,* and the gathering at the Montmor Academy suggests similar debates.

In England, informal gatherings similar to the group that Mersenne collected found an institutional home in Gresham College. Created by the will of Sir Thomas Gresham, located in his London house, and financed by income from his estate, Gresham College was an attempt to establish something of the activities of higher éducation in the city. Three of its seven professorial chairs—in medicine, geometry, and astronomy—were related to science. We know very little about its success as an educational enterprise, but we do know that English scientists quickly learned to gather there.

Two accounts set down much later by John Wallis, a prominent mathematician and physicist, tell of particular meetings which began in 1645. A group of ten men in London met regularly to consider questions of natural philosophy. After history agreed to call the group the "Invisible College," from a phrase that appears in a letter of Robert Boyle, someone was unkind enough to point out that Boyle was clearly referring to a different group. Suffice it to say that Wallis' circle continued to meet in London until the Parliamentary victory; the consequent reorganization of royalist Oxford sent a number of them to the university in 1649. John Wilkins became the Warden of Wadham College at Oxford during the Interregnum, and for a period of about ten years, Wadham College and Oxford witnessed the most intense scientific activity carried on in England. When the Restoration of the Stuart dynasty in 1660 completed the dispersion of the Oxford circle and led the majority of them to relocate in London, a group of some thirty men, actively interested in science and well known to each other, existed. After a lecture by the then Gresham professor of Astronomy, Christopher Wren, they foregathered at a nearby tavern and decided to organize formally. Two years later they took the name Royal Society.

Early in the decade, there were also significant political developments in France which directed attention to the formation of the Royal Society. With the death of Cardinal Mazarin in 1661, Louis XIV announced that henceforth he would be his own first minister. Members of the Montmor Academy began to dream of royal patronage similar to that they fancied Charles II was lavishing on his society. Indeed, the Montmor Academy was declining rapidly toward extinction as internal factions divided it. In this situation in 1663, Samuel Sorbière, who had visited England and been admitted to the Royal Society, composed a memorandum explaining the need for governmental support. Three years later, 1666, the *Académie royale des sciences* (Royal Academy of Science) was officially established through the agency of Jean Baptiste Colbert, Louis XIV's minister of finance.

From the beginning, the *Académie des sciences* was a different sort of society from the Montmor Academy. Limited to sixteen members, it attempted to bring together the leaders of science, not to propagandize the educated public, but to get on with the work of research. It was not even limited to the leaders of French science. From the Netherlands, Christiaan Huygens was brought to Paris; from Denmark, the astronomer Roemer; from Italy, Cassini—a sort of early modern brain drain. The French government, which appointed the members, also paid their salaries, and the *Académie* was its creature.

As one consequence, the *Académie* commanded a relatively full purse. Its scientists were the best equipped in Europe and in a position to carry out projects impossible to others. The *Académie* sponsored the measurement of the length of one minute of arc on the earth's surface, thus determining the size of the earth with an accuracy far beyond any earlier measurement. An expedition to South America helped to determine the distance of Mars from the earth and by indirection the dimensions of the solar system. Projects of similar scope were beyond the means of individual scientists, and the *Académie des sciences* put the entire scientific community under obligation by conducting them.

For the benefits a price was paid. If the *Académie* was financed by the French government, it was also commanded. It functioned as a sort of patent bureau for the government, frittering away the time of leading scientists on trivia. The *Académie* itself sometimes operated as a corporate body, and imposed activities on men who might have occupied themselves otherwise if left alone. Its impact on French science is difficult to assess and even to separate from other influences. This much is clear, however. The *Académie* was organized in response to the expressed feeling that the welfare of French science required governmental support. When it was organized, France led the march of European science. Thirty years later, leadership had passed unmistakably to England. Evidence does not exist to demonstrate that the organization of the *Académie* caused the relative decline. We can say that it failed to maintain the leadership that French science had won before its establishment.

In England meanwhile, the scientific community organized quite a different body, the Royal Society. Founded spontaneously by the group which gathered to hear a lecture at Gresham College, the society was a typical expression of the English genius for self-government. With prominent courtiers in its ranks, it sought royal patronage and in due course was privileged with the adjective "Royal." Charles II's purse was always insufficient to the demands made upon it. Faced with the choice of supporting Nell Gwyn or science, he never wavered in his duty, and the adjective "Royal" was by far the most valuable contribution that the Royal Society received from him. Despite its name, and despite the official sanction given its charter, it was a private organization in the fullest sense.

As one consequence, the Royal Society had a membership utterly different from that of the *Académie*. Whereas the French society attempted to assemble a scientific elite, the Royal Society opened its doors to anyone who professed an interest and rapidly filled up with chatty amateurs. Apparently everyone proposed for membership during the 17th century was elected, and joining the Royal Society was one of the fads of Restoration society. During the 1660s, the society thrived on its first flush of enthusiasm. Ten years later, its sheer amateurism nearly carried it to oblivion.

The Royal Society did not disappear, however; it is today the oldest scientific society extant. It survived in part through its good fortune in the men who served it. In its Curator of Experiments, Robert Hooke, it had a man of universal ability who could supply a modicum of scientific content to the meetings even when the members preferred to discuss two-headed calves. The society was served as well by Henry Oldenburg (c. 1620–1677), a displaced German who became the corresponding secretary. Through his correspondence, not only the scientific community in England, but a broader international one found a strand of cohesion. By founding the *Philosophical Transactions,* similar to the society in being the oldest scientific journal extant, Oldenburg institutionalized his own function and helped to create the new literary form that modern science has fostered. Through him and through the *Philosophical Transactions,* the Dutch microscopist, van Leeuwenhoek, and the Italian, Malpighi, communicated their discoveries. Through him and the *Philosophical Transactions,* Isaac Newton overcame his apprehensions and established contact with the rest of the scientific world.

To say that Hooke and Oldenburg saved the Royal Society is to speak only half the truth, however. Oldenburg died in 1677, and within ten years Hooke had ceased to be active. The functions they had performed so well were taken up by others. Perhaps nothing else indicates so clearly why the Royal Society did in fact survive—and why other societies also came into existence, survived, and grew. The scientific societies were called into being by the need for communication among scientists; the need transcended any individual man. Early in the 17th century, the survival of groups depended on individuals such as Mersenne. By the end of the century such was no longer the case, and societies patterned on the Royal Society and the *Académie* were beginning to spring up in the various corners of Europe.

There is reason to think that the informal and frequently chaotic Royal Society answered the needs of 17th century science more adequately than the better financed but rigidly structured *Académie.* Expensive projects such as measuring the earth were beyond the resources of the Royal Society, but projects of this sort were confined largely to the measurement of constants. The progress of quantitative science demanded their results, but the measurements themselves scarcely represented major steps in scientific understanding. Meanwhile, if the Royal Society could not finance such projects, it could encourage work that was far more significant in the end. The crucial word is "encourage." With its loose structure, the society could not presume to dictate or to dominate the work of its members. By its mere existence and its avowed interest, it could gently encourage, and by such means it aided the publication of one of the greatest microscopists, Robert Hooke, of one of the greatest naturalists, John Ray, of the greatest physicist, Isaac Newton, and, to a lesser extent, of the greatest chemist, Robert Boyle. The *Académie* could not begin to claim as much.

If the formation of societies is an indication of the waxing strength of the scientific movement, the invention of instruments is another. To the 17th century we owe a considerable number of the tools which have been basic to scientific research even since. At the beginning of the century, the invention of the telescope gave astronomy the instrument which revolutionized the study of the heavens. Within a decade, the microscope, which did as much for biology at a later date, appeared beside it. The first precision clock made it possible to measure time with undreamed-of accuracy. The thermometer brought temperature within the boundaries of measurement as well, although a standardized scale making possible the comparison of one thermometer with another was not devised until the 18th century. Atmospheric pressure was subjected to measurement by the barometer, and with the airpump science gained the power to vary the pressure and to create a virtual vacuum for laboratory use. No century before had contributed so richly to the instruments of research.

Beside the tangible instruments must be placed an intangible one of greater importance—the experimental method. Part of the wholesale rejection of earlier philosophies of nature, which was a central feature of the scientific revolution, was the feeling of disillusionment with earlier methods. After centuries of investigation, nothing solid had been established. The method of investigation must therefore have been wrong. The very number of men who devoted attention to method during the century is an indication of how widespread the feeling was. Bacon's *Novum Organum* was followed by Descartes' *Discourse on Method,* and Pascal, Gassendi and Newton, to name no others, all wrote on the question at greater or lesser length.

There is something unsatisfactory about all of the discussions. Bacon (1561–1626) enjoys a considerable popular reputation as the originator of the experimental method, but the reputation is not apt to survive a careful reading of the *Novum Organum** (1620). Whatever the value of his insistence on the direct examination of nature, his call for a universal natural history as the necessary foundation of science reflects the general tone of unguided observation which dominates his work. Descartes on the other hand, argued that experimentation is relevant only to the details of science, whereas reason alone can establish the general principles of natural philosophy. His confident assertion of the power of reason to probe the limits of nature exerted considerable influence on 17th and 18th century thought, but did very little to shape what we call the scientific method. Pascal's brief essays on method attempted to relate experimentation more closely to the Cartesian program, but the essays remained incomplete.

---

* *New Organon. Organon* was the name given to the corpus of Aristotle's treatises on logic.

Perhaps the best statement of the experimental method during the century is found in a manuscript by Robert Boyle. Brief but expressive, it is found in a list of the characteristics of an excellent hypothesis.

"That it enable a skilfull Naturalist to foretell future Phenomena, by their Congruity or Incongruity to it; and especially the Events of such Experiments as are aptly devised to Examine it; as Things that ought or ought not to be Consequent to it."

Most of the writings on method in the 17th century were concerned with the question of confirmation. Boyle's short statement expresses the activity of investigation which distinguishes the experimental method of modern science from logic.

Unlike the three laws of planetary motion or the sine law of refraction, the experimental method was not a specific discovery which can be ascribed to 17th century science alone. There is, of course, no single experimental method, and one can talk about it only in the most general terms to indicate a type of investigatory procedure that can be distinguished from others, such as historical research or logical inquiry. Moreover, precursors and prior examples of experimental investigation abound. Instances of it can be found in Galen's writing on physiology. Something very like the hypothetico-deductive system was examined at considerable length by the medieval school deriving from Robert Grosseteste and by the logicians at the University of Padua in the 16th century. Nevertheless, it was in the 17th century that the experimental method, the active questioning of nature under conditions defined by the experimenter, as opposed to bare observation of the phenomena that nature spontaneously presents, became a widely employed tool of scientific investigation. Whatever precursors may be legitimately pointed out, most of the early classics of experimental investigation derive from the 17th century.

The very simplicity of William Harvey's experiments in physiology illuminates the essential aspects of an experimental approach. When he cut off the circulation to his arm with a ligature and observed the changes which followed, he was imposing on nature a set of artificial conditions dictated by his question. In a similar way, the first barometer was an experiment in which a carefully defined question led Torricelli to fill a glass tube with mercury and erect it in a dish. Without the design of the experimenter, the phenomenon Torricelli observed would never have occurred. One of the best experimental enquiries of the 17th century was Newton's series of experiments on the origin of colors. It is difficult to speak of natural phenomena in relation to them. Newton contrived a set of artificial conditions under which the intent of the experimenter wholly defined the question put to nature. To the answer he had to acquiesce, of course, but the design of the experiment determined that nature had no choice but to answer "yes" or "no."

By the end of the 17th century, the scientific revolution had forged an instrument of investigation that it has wielded ever since. A good part of its success lay in developing a method adequate to its needs, and since that time the example of its success has led to ever broadening areas of imitation.

During the century, the influence of the experimental method of investigation was scarcely felt outside the domain of natural science, except for the questions it was beginning to raise in epistemology. Closely allied to method, however, was the issue of authority, and on this question the scientific revolution played a leading role in reshaping basic attitudes of western thought. European civilization had sprung up in the Middle Ages in the shadow of ancient civilization, and an awesome weight of authority had hung over it from its beginnings. On the one hand, the inspired Scriptures and the inspired Church proclaimed the will of God in matters spiritual. On the other hand, the secular legacy of ancient civilization, which was scarcely less imposing, suggested an attitude of deference toward an achievement obviously beyond the capacity of contemporary man. The Protestant churches of the Reformation accepted the Bible without question as the word of God, and Renaissance culture before it submitted eagerly to the yoke of ancient authority. Whereas Luther thought to refute Copernicus by quoting the Scriptures, an Italian humanist advised a young man to spend two years reading nothing but Cicero and to eliminate from his vocabulary any word he did not find there. Willingness to accept authority, the assumption that there must be authority more likely to be correct than the individual, continued to be a prevailing attitude well into the 17th century. Galileo railed at the peripatetics who accepted Aristotle's word in the face of sense and reason, and the words of Alexander Ross, an intelligent Scottish Aristotelian, suggest that Galileo was not dueling with a straw man.

"I follow the conduct of the most and wisest Philosophers [he replied to John Wilkins' defense of Copernican astronomy], so that I am not alone; and better it is to go astray with the best then with the worst, with company then alone."

It was an attitude not conducive to scientific enterprise, an attitude that could hardly survive unchallenged in an age of scientific revolution.

The attitude of deference had dissolved away by the end of the 17th century, and European thought was advancing rapidly toward the full-blooded optimism that characterized the Enlightenment. Undoubtedly many factors, such as economic growth and political stability, contributed to the change. Undoubtedly the change itself became a factor in the progress of science. At the same time, it appears that the success of science played a significant role in reversing the prevailing attitude. The book *Plus Ultra,* published in 1668 by Joseph Glanvill, offers an example. An episode in the so-called battle of the books, the debate on the relative attainments of the ancients and the moderns, *Plus Ultra* relied primarily upon science to argue that modern achievements had surpassed the ancients'. The title itself implies the content. A conscious play on the ancient myth that the Pillars of Hercules at the straits of Gibraltar bore the motto *"ne plus ultra"* (go no further), Glanvill's title, *Further Yet,* proclaimed that the narrow limits of the ancient intellectual world had been torn asunder. *Plus Ultra* is a catalogue of modern achievements, primarily scientific discoveries. Glanvill listed accomplishments in anatomy, mathematics, astronomy, optics, and chemistry. He cited inventions—the microscope, telescope, barometer, thermometer, air-pump. The whole tone of the work expresses the fact that authority did not command his allegiance.

The Bible continued to occupy a special place as the inspired word of God but even its authority was no longer unquestioned. The first steps in modern Biblical scholarship had begun to subject its text to historical criticism. From the point of view of science, an obscure episode in the correspondence of Isaac Newton is revealing of the change taking place. In response to a letter from Thomas Burnett, Newton wrote a brief account of the creation, using the evidence of science to confirm the reliability of *Genesis.* Compared to Luther's refutation of Copernicus by quoting the Bible, the passage embodies an exact reversal of roles. Newton's letter submitted the acceptability of *Genesis* to the authority of science. Had the matter been put to Newton in these terms, he would no doubt have rejected it, but the (probably unconscious) implication of the passage cannot be missed. That far had science moved in the rejection of authority and the elevation of unaided human faculties, that is to say, the elevation of itself, to the seat of authority now vacant.

Science also contributed to a new ideal of the function of knowledge. Where knowledge had been considered an end in itself, and the quiet contemplation of truth the highest activity in which man can engage, the assertion was now made that the end of man is action and the end of knowledge utility. The name of Francis Bacon is more closely associated than any other with the new conception, and it is frequently called Baconian utilitarianism.

"The world was made for man, Hunt, not man for the world," Bacon said to his servant. He summarized his view in the phrase, "the Kingdom of Man," which is perhaps the basic idea in the entire corpus of his writings. The kingdom of man is the physical world, the domain intended for man by God, an inheritance into which he can enter only by the path of natural science. To Bacon, knowledge was power, power by which man can subject nature to his will and force her to serve his needs. In the *New Atlantis* (1627), the first scientific utopia, Bacon described Salomon's House, an organization devoted to "enlarging the bounds of Human Empire, to the effecting of all things possible." Nearly all of the research he described in Salomon's House was practical—improved orchards, improved breeds of animals, improved medicine. Bacon himself believed that practical results derive only from true theory, and he was not in any way opposed to what we call pure research. Nevertheless, the description of Salomon's House gives an accurate account of his ultimate goal. The purpose of knowledge is the relief of man's estate, the comfort and convenience of human life.

Not every scientist of the 17th century subscribed to the ideal of Baconian utilitarianism. Indeed, it was associated, not primarily with mechanical science, but with Renaissance Naturalism and natural magic, expressing its goal of dominating nature through knowledge of her occult powers. Natural magic influenced Bacon profoundly, and since he has continued to be read while natural magicians have not, we have attached the name "Baconian utilitarianism" to an attitude widely expressed before Bacon. The manifold economic and social changes that have transformed the Western world since the 17th century have operated to select and emphasize the ideal of Baconian utilitarianism. Technology has played a major role in these changes, and technology has associated itself ever more closely with natural science. To that extent the scientific movement of the 17th century contributed to shaping an ideal of the function of knowledge that has come to be almost the ethic of modern culture.

By the end of the 17th century, modern natural science had become a prominent factor on the European scene. The day of the solitary investigator, such as Copernicus pursuing his studies in the isolation of

East Prussia, had passed, and the continued growth of the scientific movement was guaranteed now by the organized societies it had created. Already its influence was being felt in other aspects of European culture, pointing toward the Enlightenment of the 18th century, when the example of science would suggest the possibility of remodeling western civilization as a whole. It is not too much to suggest that Western history since that time can be summarized in the steady expansion of the role that science has played, transforming what was originally a culture organized around Christianity into our present one centered on science. The transformation was under way already before the scientific revolution was complete.

CHAPTER VII

The Science of Mechanics

TWO DOMINANT THEMES run through 17th century science. One, which expressed itself through the mechanical philosophy, was the urge to prune all that smacked of the occult from the body of natural philosophy. Drawing its inspiration from the atomists of the ancient world, the new conception of nature set about explaining the mechanical reality that must lie behind every phenomenon. No area of science stood immune from its influence. The second theme also traced its history to an ancient source, the Pythagorean sect. Concerned with the exact mathematical description of phenomena, it animated heliocentric astronomy. During the 17th century, the science of mechanics was its principal embodiment.

The history of the modern science of mechanics has consisted of a set of elaborations on the new conception of motion enunciated by Galileo. The first elaboration came from the hands of Descartes. Whereas Galileo had been intent on the problem posed by Copernican astronomy, Descartes focused his attention on the articulation of a new philosophy of nature. Exactly this focus helped him to take a step that Galileo never succeeded in taking, and to treat all motion in the same terms. Inertial (circular) motion around a center always remained distinct in Galileo's mind from motion toward the center, which he referred to as "natural motion." Such distinctions disappeared completely from Descartes' universe. All motions, as motions, were treated in identical terms. All changes of motion were referred to the same cause, the impact of one particle of matter on another. In such a context, it was easier to question Galileo's assumption that inertial motion is circular motion around a gravitational center. Descartes concluded that every body in motion tends always to follow a rectilinear path. It traces a curve only if something diverts it. Since nature is a plenum, every body is in fact continually diverted; nevertheless inertial motion is rectilinear.

In drawing the consequences of his conclusion, Descartes attempted the first analysis of the mechanical elements of circular motion. As mechanics, the analysis had gross deficiencies, and every beginning student of physics today can do better. Unlike the beginning student today, Descartes had no precedent on which to draw. His analysis furnished the precedent on which others drew and the foundation on which they built—until today the beginning student can quickly be taught to analyze circular motion. Descartes did reach the conclusion that a body moving in a circle constantly strives to recede from the center because of the rectilinear tendency of its motion. When we whirl a stone in a sling, we can feel the pull, "because it draws and stretches the cord in attempting to move directly away from our hand." He did not even try to find a quantitative expression for the striving away from the center, but satisfied that it exists, he made it a central factor in his system of natural philosophy.

Descartes' natural philosophy also emphasized another problem in mechanics. The conscious exclusion of all that was considered to be occult confined action to the direct contact of one body with another. Hence the problem of impact was bound to assume importance for mechanical philosophers. It was not an easy problem. Galileo had examined what he called the force of percussion without notable success, and recognizing as much, he had left his discussion unpublished. Descartes' attempt to deal with impact was one of the few cases in which he tried to introduce precise quantitative mechanics into his mechanical philosophy.

He based his analysis on the conservation of the quantity of motion. By quantity of motion he meant the product of a body's size and its velocity—a conception similar to our idea of momentum, but differing from it in so far as his "size" is not identical to our "mass" and velocity in his treatment is not a vectorial quantity. Because of the immutability of God, who is the ultimate cause of motion, he reasoned, the total quantity of motion in the universe must remain constant. There is no necessity, however, that every body's quantity of motion remain constant; in impact, motion can be transferred from one body to another. Descartes considered the two bodies in impact as a unit; the sum of their motions after impact must equal that before. He was unable to think of it solely as a problem in the conservation of motion, however; as in the case of Galileo, the notion of a "force" of percussion asserted itself.

"The force with which a body acts on another body, or resists its action, consists in this alone that each thing endeavors as much as it can to remain in the same state in which it finds itself." From this premise emerged a law of impact, his third law of nature, startling in its unexpected conclusion.

"If a body in motion meets another stronger than itself, it loses none of its motion, and if it meets a weaker one which it can move, it loses as much of its motion as it gives to the other."

In a discussion of seven different cases, Descartes distinguished quantity of motion from direction. A change of direction does not involve a change in the quantity of motion. For example, let a body in motion strike a larger one at rest. The larger one, just because it is larger, endeavors more strongly to remain in its present state, and thus the smaller (that is, weaker) one cannot move it. If it cannot move the larger one out of its way, obviously it cannot continue in the same direction, and if the larger one continues at rest, the conservation of motion demands that the smaller one continue to move with the same velocity. Hence it must rebound with its motion intact but its direction reversed. Similarly, if two equal bodies move in opposite directions, one more slowly than the other, the slower (weaker) body cannot change the state of the faster (stronger) one. Nor can it merely rebound with its original motion because the faster one is moving in that direction with greater speed. The faster body must transfer half its excess speed to the slower, and the two will move together in the original direction of the faster body.

If Descartes' treatment of impact entangled itself hopelessly in his idea of the force of a body's perseverance in its state, his mechanics contained no other clear conception of force. The simplest case in dynamics as it is now accepted is a uniform acceleration produced by a uniform force. Galileo had identified free fall as a uniformly accelerated motion, but neither Galileo nor Descartes identified its cause as a force. What could such a force be? If one answered that it is an attraction, the spectre of occult properties raised its ugly head. Galileo had avoided the whole issue by calling free fall a "natural motion." In Descartes' world, natural motions did not exist, and he worked out a mechanism, based on centrifugal tendency, to explain why bodies called heavy fall toward the earth. Heaviness was held to be the result of the multiple impacts of tiny particles all of which have tendencies away from the earth that are greater than the centrifugal tendencies of large bodies.

In a plenum, bodies with smaller centrifugal tendencies are pushed to the center, and we call them heavy. In these terms, heaviness was shorn of occult connotations. It was shorn as well of any possible reconciliation either with Galileo's conclusion that free fall is a uniformly accelerated motion or with his conclusion that all bodies fall with the same acceleration. The inability of mechanical philosophers to consider any conception of force except the "force of a moving body" became an obstacle to the development of a mathematical dynamics and tended to confine mechanics within kinematic problems, in which motions were described without reference to the forces that cause them.

This was not true of Evangelista Torricelli, the greatest of Galileo's Italian disciples. Standing outside the precincts of the mechanical philosophy, Torricelli was able to apply a frankly dynamic set of conceptions to Galileo's kinematics. Although his dynamics differed utterly from ours, basic mathematical relations inherent in a dynamic interpretation of Galileo's results emerged at once.

Torricelli began with Galileo's problem of the force of percussion. If a weight of a thousand pounds is required to break a table, how is it possible that a body weighing one hundred pounds, which has fallen from a sufficient height, can also break the table? He answered by maintaining that the heaviness of a body is an internal principle which generates in every instant an impetus equal to the weight of the body. He used the figure of a fountain—heaviness is a fountain from which impetus or momentum continually flows. If a fountain yields one gallon of water a minute, we can collect a hundred gallons by filling a gallon jug a hundred times. Similarly with the fountain called heaviness, if we collect the momenta that flow out in several instants, we can multiply the strength of the body in question. How do we collect it? By letting the body fall. When the hundred pounds rest on a table, the resistance of the table opposes and destroys the momentum generated each instant. When the body falls, no resistance annihilates the momentum; that generated in one instant is added to that generated the instant before, and the strength of the body swells continually. Hence the weight of a hundred pounds allowed to drop far enough acquires the strength of a thousand pounds needed to break the table.

Clearly, Torricelli was using a set of conceptions that look anything but familiar. The very problem he set himself looks strange, the attempt to measure the dynamic action of impact by means of a static weight. Only in this context could he think of heaviness as a fountain pouring out a stream of momenta each equal to the body's weight. The word "instant" had a special meaning for him, an ultimate unit of time, infinitesimally small, which can be divided no more. Matter he regarded as a vessel, "a bewitched vase of Circe which serves as a receptacle of force and of the momenta of impetus. Force, then, and impetus are abstractions so subtle, quintessences so spiritual that they cannot be contained in any jar except the inmost materiality of natural bodies." Torricelli was expounding a version of the medieval mechanics of impetus, the internalized force used to account for projectile motion. In his case, impetus mechanics was applied to Galileo's kinematics, and beneath his unfamiliar expressions lie a number of the basic quantitative relations of modern dynamics. By taking a dynamic view of vertical and inclined descent, with heaviness (*gravitas*) operating as motive force, he recognized the proportionality of force and acceleration. Equally he saw that the product of a constant force and the time it operates is equal to the total momentum generated in a body falling from rest.

What is more important, he was able to apply the relation derived from free fall to other situations. Among them was percussion itself. If an impetus equal to the weight of a body is added each instant, and if instants are infinitesimally short, then the strength acquired by a body in a finite time must be infinite. Torricelli agreed. But the effect of the strength will be infinite only if it is all applied instantaneously. Such is never the case. Because of the elasticity of bodies, impacts are spread out over time, and the longer the time the less the strength exercised. Torricelli had recognized that the destruction of momentum is dynamically identical to its generation. To both he applied the implicit equation

$$Ft = \Delta mv$$

The product of a uniform force and the time it operates equals the change of momentum. He applied it as well to elastic rebound, and, what is most impressive, he used it successfully to analyze a totally different problem. Imagine a large galley and a small skiff moored twenty feet from a pier. If a man pulls the galley in, all his effort scarcely gives it any velocity, and yet the pier shakes when the galley hits it. In contrast, he can set the skiff into rapid motion at once, but its effect is almost nothing when it strikes the pier.

"If we ask how long he strained in pulling the galley, he will answer that he needed perhaps a half hour of continual effort to move that huge machine twenty feet. But to pull the little skiff he spent no more than four beats of music. The force, however, which flowed continually from the worker's arms and sinews, as from a lively fountain, did not merely vanish in smoke or fly away through the air. It would have vanished if the galley had not been able to move at all, and it would all have been extinguished by that rock or anchor which prevented the movement. Instead it was all impressed in the substance of the wood and tackle of which the ship is made, and there it conserved itself and grew, neglecting the small amount that the resistance of the water can have taken away. Will it be so marvellous then if the blow which carries the momenta accumulated during half an hour has a much greater effect than that which carries no more than the forces and momenta accumulated during four beats of music?"

We invest less poetic fancy in the analysis, but we arrive at a similar result.

One has only to compare the conceptual furniture of Torricelli's analysis with that of Descartes' to understand why a dynamic rendition of Galileo's kinematics did not readily appear among the mechanical philosophers. Torricelli's lectures were in fact not published until the 18th century, but it is difficult to believe that they would have been well received had they been published at the time they were delivered. Meanwhile, in his *Opera geometrica** (1644), which did appear, he advanced another idea that exercised considerable influence on the science of mechanics.

"Two heavy bodies joined together are not able to put themselves in motion unless their common center of gravity descends."

A balance is in equilibrium if the motion of its arms does not lower the common center of gravity. When two bodies are connected by a rope over a pulley, one falls, pulling the other up, only if their common center of gravity descends in the process. Torricelli grasped the fact that two bodies isolated from external influences can be treated as one body concentrated at their center of gravity. By this means, Galileo's kinematics of heavy bodies could be extended to systems of bodies. In further exploiting that idea, the science of mechanics achieved one of its major triumphs in the 17th century.

The man responsible for exploiting the insight was a Dutch scientist, Christiaan Huygens. Son of a friend of Descartes, Huygens was raised and educated as a Cartesian, but the example of Galileo in the precise

* *Geometrical Works.*

mathematical description of motion influenced him at least as much. As a young man, he scandalized his Cartesian teacher by suggesting that Descartes' rules for impact were wrong. The word "suggest" is too weak; Huygens demonstrated Descartes' error in terms of Descartes' own principles. To Descartes, rest and motion were relative terms; since there is no space apart from bodies, we can only say that a body moves or is at rest in relation to another body. His rules of impact unfortunately yielded different results for different frames of reference. A smaller body in motion rebounds from a larger body at rest with its speed intact, while the larger body suffers no change whatever. If we shift the frame of reference, however, to consider the smaller body at rest, the larger body puts it in motion, losing as much motion as it gives to the smaller body, and the two move together after impact. Obviously, the second result is inconsistent with the first if motion and rest are relative terms, as Descartes contended. Huygens accepted the relativity of motion as beyond question. The problem then was to revise the rules of impact.

For this purpose, he imagined a thought experiment such as only a Dutchman could have proposed. A boat coasts smoothly down a quiet Dutch canal, and on the boat a man performs experiments with bodies in impact. Huygens imagined the bodies to be suspended from strings which the man holds in his hands; by bringing his hands together, he causes the bodies to strike each other. We are to understand this, of course, as a particular device to eliminate irregularities such as friction and to realize Galileo's idealized motion. The strings had the advantage of allowing him to station a second man on the shore who joins hands with the first as the boat passes. Two men jointly perform one and the same experiment. The outcome of the simplest case of impact Huygens assumed—when two perfectly hard bodies (we would say perfectly elastic), equal in size, move with equal and opposite velocities, both rebound with their original speeds unchanged. Imagine the experiment to be performed on the boat which moves at the same speed as the bodies. To the man on the shore, it appears that one body is at rest before impact, that it is given a motion equal to that which the second body had before impact, and that the second body comes to rest. By adjusting the speed of the boat, Huygens was able to proceed through all the cases involving equal bodies. To deal with unequal bodies, he further assumed that whenever a body strikes a smaller one at rest, it puts it in motion and loses whatever part of its own motion it transfers to the smaller one. By means of the boat, he now reversed the state of rest and motion. From the new frame of reference, the motion lost by the large body in moving the small one appears as motion imparted to the large by the impact of the small. "A body however large," Huygens concluded, in direct opposition to Descartes, "is moved by a body however small which strikes it with any velocity whatever." Nothing could state more clearly the 17th century's conviction of the inert passivity of matter.

Huygens gave Descartes' principle of the conservation of motion a special formulation.

"When two hard bodies collide with each other, if one of them retains after impact all of the motion it had, the other one also neither loses, nor gains motion."

Once again the boat was employed to extract all the consequences that the premise entailed. Under what conditions can bodies retain all of the motion they had before impact? Huygens showed that this can happen only when the magnitudes of the bodies are inversely proportional to their velocities. But to say it can only happen under such conditions is to say also that it happens in every impact because the relativity of motion allows us in each to choose a frame of reference in which their velocities will be in such proportion to their magnitudes. In relation to their common center of gravity, the magnitudes of two bodies in impact are always inversely proportional to their velocities, and after impact the bodies separate at the identical speeds with which they approached. The center of gravity, of course, suffers no change whatever. There is, Huygens concluded, "an admirable law of nature" which appears to be valid for all impacts of all bodies.

"It is that the center of gravity of two or three or however many bodies you wish moves always, before and after their impact, uniformly in a straight line in the same direction."

That is to say, impact can be solved by the application of Torricelli's principle. Where Torricelli had applied it only to vertical motion in the case of two bodies constrained to move together, Huygens applied it as well to inertial motions of bodies not so constrained. An isolated system of bodies can be considered as a single body concentrated at their common center of gravity. From this point of view, a purely kinematic treatment of impact without any reference whatever to the force of percussion was possible. The word "force" did not appear in the title of Huygen's treatise—*On the Motion of Bodies in Percussion*.* Despite his extensive correction of Descartes, his view of impact embodied fundamental aspects of the Cartesian one. In impact, there is no dynamic action whatever; from the point of view of the center of gravity, each body has the direction of its motion changed instantaneously, but both depart from impact with their original motions unaltered.

The very basis of Descartes' treatment, and a foundation stone of his natural philosophy, however, now appeared to be incorrect. The quantity of motion is not conserved in all impacts—at least not in all frames of reference. Since Huygens followed Descartes in distinguishing direction and speed, the quantity of motion of a body always has a positive value in his mechanics, the magnitude of a body multiplied by its speed. It was a simple matter to show that in cases in which one body alone reverses its direction, the quantity of motion does not remain constant. Another quantity does remain constant in the impact of perfectly hard bodies, however. If the magnitude of each body is multiplied by the square of its velocity, the sum of the two quantities before impact always equals the sum of the two quantities after impact. To Huygens, the result of this operation, the sum of the products of magnitude multiplied by the square of velocity, was merely a number, a number which differed in value according to the frame of reference, but a number which remained constant within each frame in the impact of perfectly hard bodies. Hence it served as a substitute for the Cartesian quantity of motion which had been shown to be incorrect. Others were to find more than a mere number in the quantity thus obtained, and it has played an ever increasing role both in the science of mechanics and in natural science as a whole.

Huygens' treatise on percussion gave a complete kinematic solution of the impact of what he called perfectly hard bodies. At much the same time that he completed the work, he also took up the question of circular motion. Once again, his starting point was Descartes, but in this case he accepted Descartes' conclusion and went beyond it to a quantitative expression of a body's effort to recede from the center. Huygens even coined a name for the effort, "centrifugal force," literally "force that flees the center." The word "force" is of some interest since his solution of impact had deliberately excluded dynamic considerations. He was willing to use the word force in this context because he con-

---

* In the original Latin: *De motu corporum ex percussione*.

sidered it to be similar to static weight, and the use of "force," meaning virtually strength, was fully acceptable in statics. When you hold a string to which a weight is tied, you feel the weight pulling straight down in the direction of the string. If you hold the same weight as you stand on a rotating turntable, you feel a similar pull in a radial direction. In both cases, the pull arises from the tendency of the body to move in the direction in which it is pulling. The analogy extends further. Imagine a tangent (*BC*) drawn to the end of a radius (*AB*) and cut by the extension of a second radius (*ADC*) that makes a small angle with the first. (See Fig. 7.1.) Let the second radius be turning at a uniform rate so that a body attached to it moves uniformly along the circle. The centrifugal force of the body, Huygens said, arises from its rectilinear inertia by which it seeks at every point to leave the curve in order to move along the tangent, and the length (*DC*) of the extended radius between the curve and the tangent measures the distance it would have moved from the circle had it been free to follow its inertial tendency at the point of contact. It is known from geometry that when the angle (at *A*) between the radii is small, the distance between the circle and the tangent (*DC*) increases in proportion to the square of the length of arc (*BD*). Since the angular motion is uniform, we can take the length of arc as a measure of time. Centrifugal force, then, is a tendency to a motion by which distance would increase as the square of

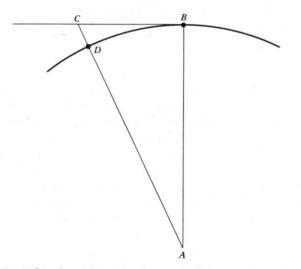

Figure 7.1.   When the angle at A remains small, DC ∝ BD².

time. As Galileo had demonstrated, weight is a tendency to a similar motion. Huygens did not ask what pulls a body back from a tangential path and holds it on a circle. Rather he accepted circular motion as given and saw centrifugal force as he saw weight, not as a force acting on a body, but as a tendency which a body has, for whatever reason, in a concrete situation. In Huygens' mind, weight and centrifugal force were more than similar phenomena; they were also complementary phenomena. Under the influence of Descartes, he held that weight is caused by a deficiency of centrifugal force. If a stone falls, an equal quantity of subtle material must move away from the earth, and the similarity of centrifugal force to weight is a causal connection.

Huygens wanted to do more than establish the connection of weight and centrifugal force. He wanted to find a formula that would express the quantity of a given centrifugal force. First, centrifugal force increases in proportion to the weight or solid matter of the body—which is as close as he came to the concept of mass. By a careful analysis of the geometry involved, he showed that it increases in proportion to the velocity squared and decreases in proportion as the diameter of the circle increases. Finally, he demonstrated that if a body moves in a given circle with a velocity equal to that which it would acquire in falling from rest through half the radius of the circle, its centrifugal force exactly equals its weight. By a simple substitution into the formula relating velocity and distance in uniformly accelerated motion, this statement yields a formula for centrifugal force that is identical to the one we use.

$$F = \frac{mv^2}{r}$$

With the dynamics of circular motion, Huygens added a weapon of significant power to the growing armory of mathematical mechanics.

He himself was the first to demonstrate how useful the formula could be by utilizing it to derive the equation for the period of a pendulum. The derivation began with a consideration of conical pendulums, a consideration in which the peculiar view of circular motion common among mechanical philosophers is clearly evident. In a conical pendulum, centrifugal force partially overcomes the gravity of the bob and holds it out from the vertical line in which it would otherwise hang. (See Fig. 7.2.) When the cord makes an angle of 45° with the vertical, it appeared intuitively that centrifugal force must equal the weight of the bob. In such a conical pendulum, the radius of the circle the

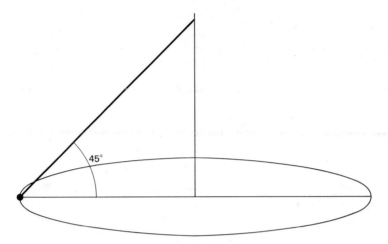

Figure 7.2.

bob describes is equal to the vertical height of the cone, and therefore (by his analysis of circular motion) the velocity of the bob is equal to that which a body would reach in falling through half the height of the cone. With this equation, he could also compare the time in which a body falls through the height of the cone to the period of the conical pendulum. He had demonstrated that all conical pendulums of the same vertical height have the same period, (See Fig. 7.3.) and that among pendulums with different vertical heights the period varies as the square root of the vertical height (AB). Galileo had shown that the period of an ordinary pendulum varies as the square root of its length, and Huygens saw that in the limiting case of a minimal oscillation, the conical pendulum becomes identical to the ordinary pendulum. The period of a conical pendulum is therefore equal to the period of an ordinary pendulum the length of which equals the vertical height of the cone (AB). By a series of simple ratios, then, employing his own analysis of the conical pendulum and Galileo's kinematics of fall, he established that the ratio between the period of a pendulum and the time of fall through its length is equal to $\pi\sqrt{2}$. But the time of fall is $\sqrt{2l/g}$. Therefore the period of a pendulum is $2\pi\sqrt{l/g}$. To Huygens, the unknown in the equation was the acceleration of gravity, g. He could measure the period and the length. Since the time of Galileo, a number of men had engaged in measuring g by measuring the distance a body falls in one second. Most results set g at about 24 ft/sec^2; the Jesuit Riccioli ar-

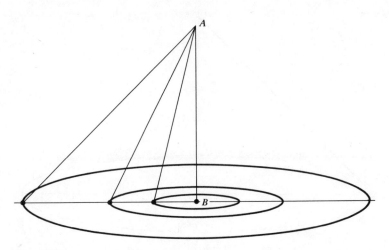

Figure 7.3.

rived at a figure of 30 ft/sec². With the pendulum, Huygens established that *g* is 32.18 ft/sec², at the latitude of the Netherlands, a figure that agrees with our best measurement today.

Beyond the results mentioned above, Huygens also demonstrated that the cycloid is the isochronous curve, the curve in which a body descends from every point to the center in equal time. (See Fig. 7.4.) Since he also demonstrated that a cycloid is the evolute of an equal cycloid, he concluded that a pendulum vibrating between two cycloidal cheeks, so that the cord wraps around the cheeks, will swing in a cycloidal path and be isochronous. On the basis of his theory, Huygens designed the first precision clock in the western world. All analyses of pendulums heretofore had been analyses of simple pendulums, the idealized case of a point mass suspended by a weightless string. Real pendulums differ. Starting with a bar swinging by one end, and imagining the bar to disintegrate into particles all of which are deflected upward, he reasoned that the common center of gravity of all the particles cannot rise higher than the point from which the center of gravity of the bar descended. (See Fig. 7.5.) From such considerations, he determined the length of a simple pendulum which vibrates with a period equal to that of the bar and hence the center of oscillation of the bar. The study of physical pendulums had begun.

In greatly extending the phenomena of motion subject to precise mathematical description, Huygens revealed himself as the heir of

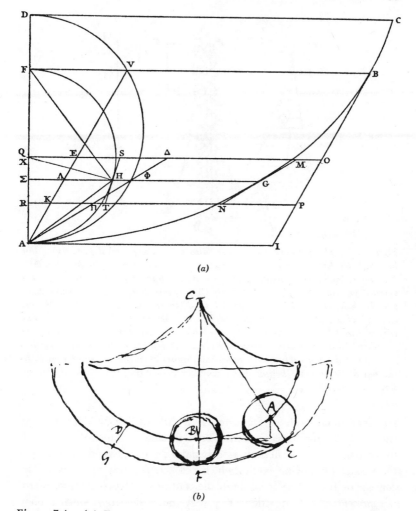

(a)

(b)

Figure 7.4.   (a) From any point B, chosen at random, along the cycloid CBGA a body will descend along the cycloid to A in the same time it takes to descend along the cycloid from C. (b) Huygens' sketch of a pendulum swinging between two cycloidal cheeks. The two curves that descend from the point C are identical cycloids. As the pendulum CA swings, the cord wraps around the cheeks and, as Huygens demonstrated, the bob as it swings follows a cycloid which is identical to those that form the cheeks.

Galileo. Between Galileo and Newton, no one contributed as much to the progress of mathematical mechanics as he. In many ways, he also

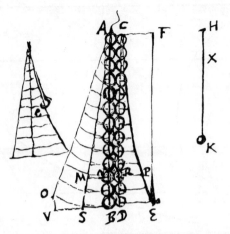

Figure 7.5. Huygens' sketch for the solution of the physical pendulum. The line of balls AB represents a solid bar that has swung from the position AO. The bar is imagined to disintegrate into its component parts when it is in the vertical position, and the line of balls CD represents the separate parts. Each part is then imagined to be deflected so as to rise straight up. The straight line AS charts the height from which each part of the bar descended. The curved line CE charts the heights to which the parts can rise when they are separated from each other. Since the center of gravity of the parts after they are separated from each other cannot rise higher than the center of gravity of the bar at its original height, the triangular area ABS must equal the curvilinear area CDE.

remained the disciple of Descartes. If it differed in detail, his universe was as rigidly mechanistic as the Cartesian, and his kinematic approach to mechanics was dictated by its demands. To Huygens, mechanics was the science of moving bodies which can interact only by impact. The concept of force appeared only in the context of circular motion, where it represented, not an action on a body, but a tendency which a body in motion has. As such, it was analogous to Descartes' "force of a body's motion," roughly what we call its momentum, a concept acceptable to mechanical philosophers.

When Huygens stood at the height of his career, a member of the French *Académie des sciences* and resident in Paris, he made the acquaintance of a brilliant young German, Gottfried Wilhelm Leibniz (1646–1716), who had come to Paris to complete his education. Huygens became Leibniz's mentor in mathematics and mechanics, and Leibniz carried some of his conclusions to a higher level of generality.

In 1686, Leibniz fluttered the philosophic dovecots of Europe by publishing an article entitled "A Brief Demonstration of a Memorable Error by Descartes."* Its central concern was the Cartesian conception of the force of a body's motion; Leibniz's conviction that Descartes' conception was wrong is obvious from the title. As the premise of his demonstration, Leibniz asserted the proposition that a body which falls from a given height acquires sufficient force to carry it back to its original height if nothing external interferes. In somewhat different terms, the principle had been current in 17th century mechanics since the time of Galileo. Ultimately, it rested on the conviction that a perpetual motion is impossible, and Leibniz insisted on this justification. If a body that falls from four feet were to acquire a force sufficient to raise it to five, then the force which causes it to rise the additional foot could be tapped off and turned to another purpose (including the overcoming of friction, which prevents the ideal rise to four feet). Something could be obtained for nothing, which is impossible. Leibniz assumed, second, that the same force necessary to lift four pounds one foot can lift one pound four feet. It was a principle which Descartes himself has used, and powerful arguments supported it. If the four pounds were divided into four units of one pound, lifting them one after another to a height of one foot would clearly be equivalent to lifting one pound one time after another through four stages of one foot each. Now set at one unit the velocity that a body acquires in falling one foot. Four pounds falling one foot must acquire a force of four units; and by Descartes' formula, that force transferred to the body of one pound will give it a velocity of four. Once again Galileo stood in the way of Descartes, for Galileo had proved that if a body projected upwards with one unit of velocity rises one foot, a body projected upwards with four units of velocity will rise sixteen feet.

"There is thus a big difference [Leibniz concluded] between motive force and quantity of motion, and the one cannot be calculated by the other, as we undertook to show. It seems from this that *force* is rather to be estimated from the quantity of the *effect* which it can produce; for example, from the height to which it can elevate a heavy body of a given magnitude and kind but not from the velocity which it can impress upon the body."

The conservation of the quantity of motion had been one of the

* In the original Latin: "Brevis demonstratio erroris memorabilis Cartesii."

pillars of Cartesian natural philosophy, and the principle of conservation itself agreed with Leibniz's view of the world. If force could ever be increased, he said, "the effect would be more powerful than the cause, or rather there would be a mechanical perpetual motion, that is a motion which could reproduce its cause and something more, which is absurd. But if the force could be diminished, it would perish at last entirely; for never being able to increase, and being able nevertheless to diminish, it would always go more and more into decay, which is without doubt contrary to the order of things." Not even the world as a whole can increase its force without a new impulse from without. Force is conserved, but what is its measure? Leibniz argued that velocity does not measure it satisfactorily. Only an effect that wholly exhausts it can measure force, and such an effect is lifting a weight. Thus Leibniz held that weight multiplied by the height it falls or rises must take precedence over mass multiplied by velocity. Descartes' mistake had been to confuse quantity of motion with motive force. Can force itself be measured apart from its effect? Leibniz's answer is implicit in the example he used. The measure of force is the mass of a body multiplied by the square of its velocity—Huygens' number.

Leibniz argued that the concept, quantity of motion, had been taken from the simple machines, such as the balance. When the arms of a balance in equilibrium are made to turn, the two weights are inversely proportional to their velocities; since they are in equilibrium, their forces appear to be equal. In such cases alone, static situations, Leibniz contended, is force equal to quantity of motion. He referred to static forces as dead forces, the beginning or the end of a tendency to motion (what he called a *conatus*). In the case of living forces, however, forces which act with a completed impetus, quantity of motion cannot be the measure. "For living power is to dead power, or impetus is to conatus, as a line is to a point or as a plane is to a line."

Obviously, living force (*vis viva*) represented much more than a mere number to Leibniz. The essence of being itself, living force carried a load of metaphysical significance with implications stretching far beyond the realm of mechanics. The conservation of *vis viva* was to Leibniz equivalent to the eternity of God's creation. Within mechanics, it offered insights on a number of problems. The first of course was the increase of force in uniformly accelerated motion, the problem he used to demonstrate Descartes' error. The second was impact, in which he drew on Huygens' analysis but went beyond it. Elasticity always presented a problem to Huygens, and he treated the ideal case as the impact of perfectly hard bodies. In such a case, impact was instantaneous, and the original motion continued with its direction reversed. Leibniz undertook to analyze the dynamics of elastic impact, arguing that the *vis viva* of the moving body is converted into elastic force as the body is brought to rest, and regenerated from the elastic force as a new motion in the opposite direction is produced. Leibniz extended his analysis to include inelastic impact. When two equal chunks of soft clay which are moving with equal velocities in opposite directions strike each other, they both come to rest. What happens to the living force in this case? Leibniz agreed that the chunks, as chunks, lose their living force.

"But this loss of the total force does not detract from the inviolable truth of the law of the conservation of the same force in the world. For that which is absorbed by the minute parts is not absolutely lost for the universe, although it is lost for the total force of the concurrent bodies."

The suggestion that the force lost by the large bodies is transferred to their parts was pregnant with meaning for the future. Leibniz himself asserted it on a priori grounds to save the principle of the conservation of living force. He did not realize that the force of the parts (let us say, their motion) can be measured as heat.

With Leibniz, Huygens' deliberate attempt to restrict mechanics to kinematics, to discuss motions without reference to force, was abandoned, and he coined the word "dynamics" to describe a mechanics built on the concept of force. Leibniz's concept of force, however, was a different concept from the one that modern physics, following Newton, employs. "Force" as Leibniz used it can be most readily translated into our term, "kinetic energy." Much as his philosophy of nature differed from Descartes', it still accepted the premise that force is not something that acts on bodies to change their motion, but something that bodies have. In his idea of dead force, Leibniz approached the other conception, but he limited dead force to static situations. Although he compared elastic force to heaviness, as a dead force from which a living force can emerge, he never carried the analysis beyond the bare words.

Leibniz's work in mechanics exploited the earlier success of Galileo and Huygens, both of whom expressed their results in the ratios of classical geometry. The limitations of geometry confined mechanics largely to problems in which uniform acceleration represented the maximum complexity. Huygens succeeded in transcending those limits in certain problems, and his demonstrations of the isochronous property of the cycloid and of the center of oscillation of the physical pendulum were among the ultimate achievements of a mechanics expressed in classical geometry. By the end of the century, however, a new mathematical tool of immense power had been invented, the infinitesimal calculus. Leibniz himself was one of the inventors. With the calculus, motions of greater complexity could be subjected to precise description with the aid of the conceptual tools that mechanics developed during the century.

Most of the major steps forward in mechanics during the century involved the contradiction of Descartes. Although the mechanical philosophy asserted that the particles of matter of which the universe is composed are governed in their motions by the laws of mechanics, the precise description of motions led repeatedly to conflict between the science of mechanics and the mechanical philosophy. In nothing was this more evident than in Galileo's description of uniformly accelerated motion; Descartes ignored it, and no successful mechanism to explain it was invented during the century. Leibniz's argument on living force rested in the end on Galileo's conception of uniformly accelerated motion. In Leibniz, the conflict between the two began to resolve itself by modifications of the mechanical philosophy. He contended that nature is mechanical only on the level of phenomena, and that ultimate reality consists of centers of activity, a conception utterly opposed to the complete passivity of matter in the mechanical philosophy. Even in Leibniz, "force" refers to the activity of a body, not to an action on a body. The development of a conception of force as action on a body to change its state of motion, a conception that contributed greatly to the further elaboration of mathematical mechanics, was inhibited by the mechanical philosophy during the century. What it might contribute to mathematical mechanics was suggested by Torricelli, but in terms the mechanical philosophy could not accept. It remained for Isaac Newton to pick up that conception again and to use it both to extend mechanics and to revise the mechanical philosophy.

CHAPTER VIII

Newtonian Dynamics

E VERYONE ACKNOWLEDGES the position of Isaac Newton in the history of science in general and in the history of 17th century science in particular. Not only was Newton's achievement monumental, so that it stands as one of the supreme accomplishments of the human intellect, but it also drew together the principal strands of 17th century science, solving major problems left unresolved by the scientific revolution. In solving problems, his work did not in any way mark an end or pause for the scientific enterprise. Like all creations of genius, his books opened two new questions for every old one they closed, and if his work summed up the scientific revolution of the 17th century, it also inaugurated 18th century physical science. In Newton, the mechanical philosophy of nature, fundamentally revised, attained a degree of sophistication such that it could furnish the framework of scientific thought in the western world for another two hundred years.

Newton occupies a special position in the history of science for other reasons as well. Since he almost never destroyed a paper—whole stacks of sheets with nothing but raw arithmetical calculations survive—our study of him is not confined to completed and polished works. From his reading he took copious notes which enables us to specify the major influences on him; and through various notebooks which reach back into his undergraduate days, we can trace the steps of his own investigation of nature. The result is a detailed picture, unique in the history of thought, of the progress of a master intellect, a picture which enables us to comprehend Newton's work as he conceived it and to place it firmly in the context of 17th century science.

That context, inevitably, was the prevailing mechanical philosophy of nature, which fostered the initial steps in scientific thought that Newton took. While he was still an undergraduate, Newton discovered the writings of the mechanical philosophers—Descartes, Gassendi, Hobbes, Boyle, and others. Forthwith, he was converted to their view. In a notebook, he jotted down passages from their works and questions they raised, and he convinced himself of the advantages of the atomist version of the mechanical philosophy. The entries in the notebook were the first installment of a lifetime's speculation on the ultimate nature of physical reality.

Before 1675, his speculations had solidified into a system of nature of his own creation. In that year, he submitted a version of it to the Royal Society under the title "An Hypothesis Explaining the Properties of Light." As the title implies, the paper concerned itself primarily with the explanation of optical phenomena, especially the periodic phenomena of "Newton's rings" which he described in an accompanying manuscript. It went far beyond optics, however, and constituted a brief but fully elaborated mechanical system of nature. Basic to it was the assertion that an aether, a fluid composed of minute particles, pervades all space. Varying in density, the aether alters the direction of corpuscles of light passing through it, and as far as optical phenomena were concerned, the point of the "Hypothesis" was to show how all the phenomena of light can be explained by such changes in direction. Beyond optics, he employed the aether to explain such diverse phenomena as sensation and muscular action, the cohesion of bodies, and heaviness. He argued that all bodies are made of condensed aether, and in explaining its condensation in the sun, he included his first vague public suggestion of the law of universal gravitation. As condensation of the aether in the earth entails a continual movement of aether toward the earth, bearing gross bodies down and making them appear heavy, so its condensation in the sun sets up a similar movement by which planets are held in their orbits.

The "Hypothesis of Light" embodied all the standard features of mechanical philosophies of nature. A special feature of Newton's speculations was the role that a particular group of phenomena played in them, phenomena that mostly appeared already in his undergraduate notebook and continued to be cited in every version of his speculations until their final statement in the "Queries" attached to the *Opticks*. As a matter of course, they appeared in the "Hypothesis." One such phenomenon was the cohesion of bodies, usually ascribed in mechanical systems to the interlocking of parts, and explained by Descartes by the mere relative rest of parts. Newton was dissatisfied with both solutions of the problem. The expansion of gases suggested another problem. When Robert Boyle formulated the concept of pressure in the air, he used the analogy of a fleece. When the fleece is compressed, its hairs bend and come together; when the compressing force is removed, they resume their original position. Experiments revealed that air can expand thousands of times in volume, however, and Newton was convinced that a crude mechanical analogy like Boyle's could not explain expansions of such magnitude. Two types of chemical phenomena caught his eye. In some reactions, heat is generated. In Newton's view, heat is a sensation caused by motion of the particles of which bodies are composed; where does the motion come from when two cold substances are gently mixed together? Equally, reactions displaying affinities intrigued him. To use an example which he considered identical to such reactions, it was difficult to explain why water mixes with wine but does not mix with oil. He spoke of a "certain secret principle in nature by which liquors are sociable to some things and unsociable to others." A secret principle of sociability—the very words summoned up the ghosts of occult qualities which the mechanical philosophy had been created to exorcise. Indeed, all of the crucial phenomena on which Newton speculated throughout his life had one property in common—all of them were problem phenomena difficult to explain by the standard devices of the mechanical philosophy—the shapes, sizes, and motions of particles.

There were bound to be phenomena difficult for the mechanical philosophy, of course. The philosophy was built on the premise that the reality of nature is not identical to the appearances our senses depict. As we have seen, microscopic mechanisms were imagined to explain such difficulties away. Obviously Newton designed his aethereal hypothesis for that purpose. Nevertheless, he was clearly dissatisfied with the standard mechanistic explanations of such phenomena, and by 1686 and 1687, when he composed the *Principia,* forces between particles had replaced the earlier aether in his speculations. Twenty years later, in the first Latin edition of the *Opticks* (1706), in what we now know as Query 31, Newton gave these speculations their definitive form.

"Have not the small Particles of Bodies certain Powers, Virtues, or Forces by which they act at a distance, not only upon the Rays of Light for reflecting, refracting, and inflecting them, but also upon one another for producing a great Part of the Phaenomena of Nature? For it's well known, that Bodies act one upon another by the Attractions of Gravity, Magnetism, and Electricity; and these Instances show the Tenor and

Course of Nature, and make it not improbable but that there may be more attractive Powers than these. For Nature is very consonant and conformable to her self."

Query 31 proceeded then to detail the evidence for such an assertion. Much of the evidence was chemical. Salt of tartar (K_2CO_3) deliquesces (or, in the more picturesque phrase of the 17th century, it runs *per deliquium*); only with difficulty can it be separated from the water it thus takes up. Obviously, salt of tartar attracts water. When an acid is poured on iron filings, heat and ebullition accompany the dissolution because the mutual attraction causes the particles to rush together with violence.

"And is it not for the same reason that well rectified Spirit of Wine [alcohol] poured on the same compound Spirit [of nitre] flashes; and that the *Pulvis fulminans* [literally, lightning powder], composed of Sulphur, Nitre, and Salt of Tartar, goes off with a more sudden and violent Explosion than Gun-powder, the acid Spirits of the Sulphur and Nitre rushing towards one another, and towards the Salt of Tartar, with so great a violence, as by the shock to turn the whole at once into Vapor and Flame?"

To the reactions producing heat, he added those manifesting elective affinities such as displacement reactions in which the addition of one metal to an acid solution precipitates another. Not all the forces between particles are attractive; some particles repel each other. The solution of salts in water requires such repulsions, because the whole solution becomes salty even though the salt is heavier than water and would sink to the bottom if its particles did not repel each other. Non-chemical phenomena point to the same forces. The cohesion of bodies and capillary action manifest attractions, whereas the expansion of gases is the product of repulsions. What is the relation of attractions and repulsions? In algebra, Newton said, negatives begin where positives end. In a similar way, particles of matter attract each other strongly at very close quarters, causing bodies to cohere. If a particle is somehow shaken loose, however, and recedes beyond a certain distance, a repulsion replaces the attraction so that water vapor, for example, expands to enormous volumes.

Newton's admission of forces acting between particles of matter constituted a major break with the prevailing mechanical philosophy of nature. His treatment of magnetism offers an instructive example of the change. In the 16th century, magnetism was the foremost example of the mysterious influences thought to pervade the universe. Correspondingly, mechanical philosophers had felt compelled to explain magnetic attraction away by inventing an invisible mechanism to account for it. Newton had done the same in a youthful writing. In his mature works, magnetic attraction was presented as an example of forces that act at a distance. Newton also approached earlier patterns of thought in the specificity attached to his forces. In Query 31, when he discussed attractions and repulsions between particles, he did not mean one universal force whereby all particles attract those near and repel those at a distance. In cases of chemical affinities, for example, certain substances attract only certain others. The repulsive forces that disperse dissolved salts were held to operate only between the particles of salt, not between the salt and the water. It is small wonder that Newton's critics felt he was reverting to the style of Renaissance Naturalism and undermining the very foundation on which science rested.

Newton himself considered forces between particles, not as a denial of the mechanical philosophy, but as the conception needed to perfect it. By adding a third category, force, to matter and motion, he sought to reconcile mathematical mechanics to the mechanical philosophy. Force to him was never an obscure qualitative action, as the sympathies and antipathies of Renaissance Naturalism had been. He set it in a precise mechanical context in which force was measured by the quantity of motion it could generate. It is true that he never succeeded in reducing most of the forces discussed in Query 31 to a mathematical description. In an interesting experiment with a drop of orange juice between two sheets of glass, he tried to quantify the forces in capillary action. (See Fig. 8.1.) By measuring the distance between the sheets of glass and the area of contact with the orange juice, he computed the attraction in terms of the weight of juice being lifted. In the *Principia,*

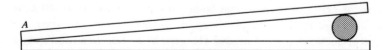

Figure 8.1.   Measurement of capillary forces. The two sheets of glass were about two feet long, touching at one end, separated at the other by a distance small enough that the drop of orange juice touches both pieces. When the end A is raised, the weight of the orange juice acts against the capillary attraction, and Newton sought to measure the capillary attraction by bringing the two forces into equilibrium.

he demonstrated that Boyle's Law must follow if particles of air repel each other with a force inversely proportional to the distance between them. Most of the forces mentioned in Query 31 were discussed only in qualitative terms, which merely pointed out the evidence that appeared to demonstrate their existence. In principle, however, all of them were subject to exact mathematical description. Thus to Newton the concept of force represented the means by which the Galilean tradition could be introduced into the mechanical philosophy. And with one force he succeeded in carrying the work to its full and magnificent conclusion. Without the concept of force, the law of universal gravitation was inconceivable. In the law of universal gravitation, the concept of force carried natural science to a new level of sophistication that has stood ever since as the paradigm of a scientific demonstration.

Newton's interest in mechanics derived from his earliest steps in natural philosophy. One of the headings in his undergraduate notebook was "Violent Motion," and under it he entered a short essay on projectiles that contained an approach to the principle of inertia. Before 1664 was out, he had done more than approach the principle of inertia.

"Every thing doth naturally persevere in that state in which it is unless it be interrupted by some external cause, [he declared], hence a body once moved will always keep the same celerity, quantity, and determination of its motion."

The specific language of the statement betrays the influence of Descartes, whose *Principles of Philosophy* Newton had been reading, and the set of propositions in which Newton included it suggests the mechanical philosophy of nature. Above the propositions, Newton wrote the title "Of Reflections," which is to say he was considering the problem of impact, the sole mode of action in the accepted mechanical conception of nature. Before he laid the subject down, he had arrived at the conclusion to which Huygens attained some five years earlier—that the center of gravity of two isolated bodies in impact continues at rest or in uniform motion in a straight line. In another paper from about the same time, Newton went beyond Huygens by adding rotational motion to bodies in impact and arriving at the principle of the conservation of angular momentum. Under the heading, "The Laws of Motion," he derived a general formula for the impact of any two bodies with both translational and angular motions. The title of the paper again expresses the broader context in which it fits. In the 1660s, the laws of motion meant the law of impact to Newton.

Nevertheless, a different enquiry was beginning to assert itself. In the propositions "Of Reflections," Newton considered the motion of bodies differing in size.

"Hence it appears how and why amongst bodies moved some require a more potent or efficacious cause, others a less, to hinder or help their velocity. And the power of this cause is usually called force. And as this cause useth or applyeth its power or force to hinder or change the perseverance of bodies in their state, it is said to Indeavour to change their perseverance."

What is force? In the content of the prevailing mechanical philosophy, it could mean only one thing: "Force is the pressure or crowding of one body upon another." With the statement neither Descartes nor Gassendi nor Boyle would have disagreed. Nevertheless, Newton was posing a question they had not put. Descartes, who tended to think of the moving particle as a causal agent, had spoken of the "force of a body's motion." On the other hand, Newton was thinking in terms of an abstract quantity which could measure the change in the motion of a moving body. Impact was the sole origin of force which he was then prepared to admit, so that "force" as he used it did not differ ontologically from Descartes' "force of a moving body."

"If two bodies p and r meet the one the other, the resistance in both is the same for so much as p presseth upon r so much r presseth on p. And therefore they must both suffer an equal mutation in their motion."

The proposition had a precedent in Descartes' assertion that one body in impact can gain only so much motion as another loses, but again Newton's statement was firmly embedded in a context of mathematical mechanics that looked back to Galileo as much as to Descartes. In the concept of force, as that which generates a change of motion, lay the kernel of his contribution to mechanics.

In another early paper, Newton took up the problem of circular motion. Profiting from Descartes' insight, he seized the basic physical elements of circular motion—a body must be diverted continually from its natural rectilinear course in order to follow a circular path. The obvious extension of the point of view adopted on impact should have led Newton to investigate the force which deflects such a body into a circle. He did not take that path, however, but like the other early students of circular motion, he looked at the tendency away from the center exerted by a body constrained to circular motion—Huygens' centrifugal force. Like Huygens, he too sought its quantitative measure, a difficult problem with a concept of force measuring the total change in motion that occurs in impact. To utilize the concept in circular motion, he imagined the moving body to strike an infinite number of identical bodies as it is deflected around a circle, while all of the motion transmitted to the other bodies is transferred to and concentrated in one of them. (See Fig. 8.2.) In this way, he arrived at the idea of the total force exerted

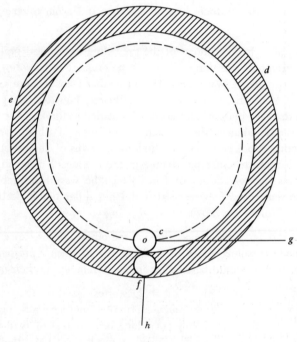

(a)

Figure 8.2.   (a) The cylindrical body def *constrains the body* o *to move in a circular path. When* o *is at* c, *it tends to move along the line* cg *and presses against the cylinder. Imagine* def *to be composed of a number of separate bodies such as that at* f. *In moving around the circle, the body* o *presses against each of them, imparting motion to it. Newton imagined all of that motion to be transferred to* f, *and its motion along* fh *constituted a measure of the total force of the body* o *away from the center in the course of one full revolution. (b) Newton's quantitative treatment of circular motion. In this figure the body (at* b) *follows a rectangular course* abcd *inside the cylindrical body. Newton showed that the force at the four reflections is to the force of the body's motion as the length of its path (*ab + bc + cd + da*) is to the radius* nb. *He further demonstrated that when the square is changed to a polygon the same ratio holds, until, as the polygon approaches the circle as a limit the ratio becomes that of the circumference to the radius.*

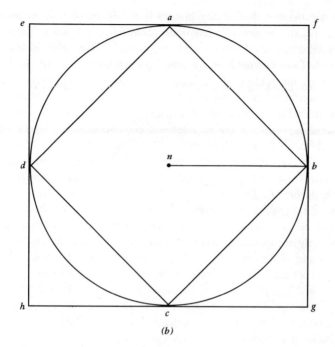

(b)

by the body in one revolution (equal to the total motion generated in the other body), an idea similar to the total force of gravity exerted on a body while it falls, say, for one minute, and also similar to the force of an impact. By a geometrical analysis starting with a square path and approaching a circle as the limiting polygon with an infinite number of sides, he demonstrated that the total force in one revolution is to the force of the body's motion (its momentum in our terms) as the circumference of the circle is to its radius. (See Fig. 8.2.) If we divide the total force by the time necessary for one revolution $\frac{2\pi r}{v}$, we can reduce Newton's answer to our formula for centrifugal force $(F = \frac{mv^2}{r})$.

In another paper from the 1660s, Newton used this formula to compare the centrifugal tendency of the moon to the acceleration of gravity on the surface of the earth, and to compare the centrifugal tendencies of the planets to each other. The latter problem was merely a matter of substituting his formula for centrifugal force into Kepler's third law, with the assumption that the planets travel in perfect circles, and he found that the tendency to recede decreases in proportion to the square of the orbit's radius. In the case of the moon, he found that its tendency to recede from the earth is one four thousandth of the acceleration of gravity on the surface of the earth, a figure which approximated the inverse square relation since he was placing the moon at a distance sixty times the radius of the earth. The paper contains the basic quantitative relations on which the law of universal gravitation rests.

Much later, Newton said that in 1666 he calculated to see if the force of gravity extends to the moon and holds it in its orbit, and he found the figures to answer "pretty nearly." Obviously, he was referring to this document. Two points must be underscored, however. The law of universal gravitation demands an exact correlation of the measured acceleration of gravity with the acceleration of the moon. Newton had found only an approximation. He used the figure for the radius of the earth found in Galileo's *Dialogue,* a figure which was too low, and only later did an accurate measurement of the earth give him the power to correct it. Meanwhile the correlation was not exact. Second, the paper did not use the concept of attraction at all. Still thinking within the framework of the prevailing mechanical philosophy, he spoke, not of gravitational attraction, but of tendency to recede.

An interlude of more than ten years interrupted Newton's study of mechanics as optics and mathematics commanded his attention. In 1679, he received a letter from Robert Hooke, now secretary of the Royal Society after the death of Henry Oldenburg, asking Newton to resume his philosophical correspondence. In his reply, Newton refused to enter into a regular correspondence. He had "shook hands with Philosophy," and grudged the time spent on it. He could not quite leave it at that, however, and to fill out the letter, he suggested an experiment to prove the rotation of the earth. (See Fig. 8.3.) The old argument against the motion of the earth held that a body dropped from a high tower should fall to the west as the earth turns by its vertical path; Newton suggested in contrast that such a body should fall to the east because its initial tangential velocity at the top of the tower exceeds that of the tower's foot. A diagram showed the path of the body as part of a spiral ending at the center of the earth. This was a slip, and Hooke, who had suffered one public humiliation at Newton's hands, was unable to let it pass. The trajectory of a body imagined to fall through the earth without resistance would not end at the center; rather the path would be a sort of ellipse with the body returning to its original height. The conclusion

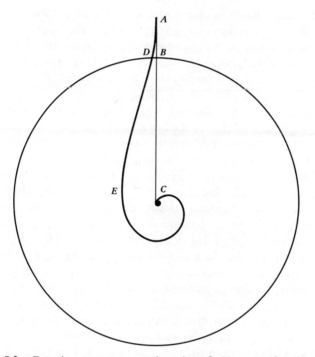

Figure 8.3. Experiment to prove motion of earth. Newton's drawing of the path of a falling body when released from the top A *of the tower* AB *on an earth that turns on its axis.*

followed, he said, from his theory of orbital motion involving a tangential motion plus a central attraction. Newton was not the man to take corrections kindly. His reply was dry and crisp as a piece of burned bacon. Acknowledging his error, he went on to correct Hooke's description of the orbit, which could not, he said, be a closed ellipse. Hooke's reply contained a second bombshell. If the central attraction were constant, Newton's suggested orbit would be correct; but he assumed that the attraction decreases in proportion to the square of the distance. Newton did not reply a second time, but he later acknowledged that Hooke's letter stimulated him to demonstrate that when a body revolves in an elliptical orbit around a center of attraction placed at one focus, the force of attraction must vary inversely as the square of the distance from the focus. Thus in 1679 or 1680, Newton demonstrated one of the two central propositions on which the law of universal gravitation rested.

Why was he able to proceed further in 1679 than in 1666? When the *Principia* was being completed in 1686, Hooke claimed that Newton had plagiarized from him. Almost universally, historians have rejected the claim, and the early papers of Newton cited above (papers of which Hooke knew nothing, of course) show how far he had advanced before the correspondence with Hooke. Moreover, with Hooke, gravitation was always an idea discussed verbally but not demonstrated mathematically, whereas the validity of the law of universal gravitation depended entirely on the mathematical demonstrations, which Newton alone supplied. In 1666, however, Newton did not think of central attractions, but of centrifugal tendencies. Hooke was the man who stood the inverted problem right side up, identified the mechanical elements of orbital motion as a tangential velocity and a central attraction, and thus put the question into a form from which the concept of universal gravitation could emerge. Let us add that Hooke's seed fell on prepared soil in which it could take root and sprout. The timing could not have been better. Hooke's suggestion of a central attraction came exactly at the time when Newton's speculations had led him to assert the existence of forces between particles. He was in a position to accept the idea of attraction as he had not been before. The idea of attraction, in turn, offered physical content to the mathematical abstraction of force toward which his earlier work in mechanics had moved. In a word, all of the factors were now present to produce the concept of universal gravitation.

But nothing came from the correspondence with Hooke except a private manuscript demonstrating that elliptical orbits could result from an inverse square attraction. In 1679, Newton was recuperating from an emotional breakdown; for five years, he virtually isolated himself from the scientific world outside Cambridge. In August of 1684, he received a visit from Edmond Halley, who had been pondering the orbital problem without success. Point-blank, Halley asked what path a body which orbits another attracting it with a force inversely proportional to the square of the distance would follow. An ellipse, Newton replied. How do you know? Why, I have computed it. When he went to look for the paper, however, he could not find it! Before long, he had demonstrated the proposition anew, and the ultimate result of that meeting was the *Philosophiae naturalis principia mathematica*,* the imperishable monument that ensures Newton's position in the history of

* *The Mathematical Principles of Natural Philosophy.*

science. Before he had left Cambridge, Halley had won a promise from Newton that he would send him the demonstration. What he received later that fall, and laid before the Royal Society, was a short tract on motion which contained key propositions of the ultimate work. With the encouragement of the society, Newton went on to complete it, and it appeared in July, 1687. History has agreed that without Halley, who not only encouraged Newton but financed the publication from his own slender resources, the *Principia* would not have been written. Perhaps the judgment is correct, but other factors were also involved. The Newton Halley approached in 1684 was a man five years removed from the depression of 1679 and open to external stimulation as he had not been earlier. Hooke's letter in 1679 arrived unplanned at an opportune point in Newton's intellectual development; Halley's visit in 1684 came equally unplanned at a happy point in his emotional life. In the spring of 1684, Newton had started a treatise on mathematics, suggesting that he was beginning again to look outward toward the scientific community. Halley received a short tract on motion in December 1684, but Newton was already engaged in extensive revisions that greatly increased its size and its shape. Halley may have tapped the fountain, but once it was tapped, the *Principia* flowed spontaneously and unhindered from the inexhaustible reservoir of Newton's genius.

Book One of the *Principia* says nothing about universal gravitation. It is a treatise on rational mechanics, which prepared the ground for the integration of orbital motion into a unified system of mechanics embracing alike terrestrial and celestial phenomena. The importance of the *Principia* lies far more in the first book than in the law of universal gravitation. In Book One, Newton brought the 17th century science of mechanics to its highest level of perfection, placing it in the position it has occupied ever since as the recognized model of a successfully mathematized science.

The book opens with basic definitions and three laws of motion. The first law states the principle of inertia in the form that is still employed, but the concept itself stemmed directly from Galileo and Descartes. The third law, the principle of action and reaction, was original with Newton, but it can be seen as an extension, in terms of dynamics, of the changes of motion in impact, which Huygens had demonstrated earlier. On the other hand, the second law and its associated definitions effectively introduced the concept of force into rational mechanics. With the concept of force, the kinematics of Galileo could be completed by the science of dynamics. "The change of motion is proportional to the motive force impressed; and is made in the direction of the right line in which that force is impressed." Strictly interpreted, Newton's words say that $F = \Delta mv$, not $F = ma$, or $F = \dfrac{d}{dt} mv$, the forms of the second law with which we are familiar. Newton's statement of the law reflects at once its source in his early consideration of impact and the demands of the geometry in which he presented the *Principia*. He considered that $F = \Delta mv$ approaches $F = ma$ as a limit when Δt approaches zero. Involved in the definition of force was the definition of mass, now clearly distinguished from weight for the first time.

The laws of motion in the *Principia* must be compared with his early paper, "The Laws of Motion." In the early paper, the laws were summed up in a generalized formula on impact. In the *Principia,* he dismissed impact in two corollaries to the laws, which treated it as a special case of inertial motion. His attention was riveted now on the motion of bodies under the influence of divers forces.

Book One is concerned to apply the laws of motion to point masses, particularly to point masses orbiting attracting centers. For this purpose, Newton coined the term "centripetal force," force that seeks the center, in deliberate contrast to Huygens' term, "centrifugal force." The phrase repeats the insight of Hooke, announced to Newton in the correspondence of 1679, and in so far as the primary advance that Newton's treatment of circular motion made over Huygens' is embodied in the point of view that the phrase repeats, Hooke's contribution cannot be ignored. When Newton went on to cover its bones with the flesh of mathematical demonstration, he stepped onto ground that Hooke never approached. Newton demonstrated that Kepler's three laws of planetary motion can be derived from dynamics. The law of areas must hold in all cases in which a moving body is deflected from its inertial path by an attracting force. When such a force varies in strength inversely as the square of the distance, bodies will orbit in one of the conics—an ellipse (or its limit, a circle) when the tangential velocity is below a critical value. In the case of an inverse square force, moreover, a number of bodies orbiting a single attracting center must obey Kepler's third law. The inverse square relation had been derived initially, of course, from the substitution of the law of centripetal force into Kepler's third law. The demonstration that Kepler's first law, the elliptical orbit, also follows from an inverse square force was immensely difficult. It was one of the key propositions on which the law of universal gravitation stood.

Whereas Book One devoted itself to idealized problems of point masses moving without friction, Book Two considered bodies moving through resisting fluids and the movements of such fluids themselves. Book One rested on the earlier achievements of Galileo, Descartes, and Huygens, which it raised to a higher level of sophistication. In the case of Book Two, only the crudest precedents were available, so that it constitutes the effective beginning of mathematical fluid dynamics. Inevitably the pioneering work contained errors, but the achievement of bringing a whole new range of problems within the scope of rational mechanics was not less impressive than the achievement of Book One. As its climax, Book Two turned to the examination of Descartes' vortices. Newton demonstrated that a vortex can never yield a system of planets moving according to Kepler's three laws. What was even more compelling, he proved that a vortex cannot be a self-sustaining system, but continues in uniform motion only as long as an external force continues to turn its central body. As he later expressed it, the system of vortices is pressed with many difficulties.

With the ground prepared in Book One and the Cartesian system demolished in Book Two, Newton turned in Book Three to the application of his dynamics to the system of the world. Astronomy presented two systems of a central body circled by satellites that obey Kepler's third law—the solar system and Jupiter with its moons. By invoking the principle of uniformity, he concluded that the inverse square forces in operation must be identical in nature. As good fortune has it, a satellite also circles the earth, but in this case, of course, only one. Even had there been two, conforming to Kepler's third law, Newton's demonstration would have remained incomplete. His purpose was to prove, not only that the forces that hold divers satellites in their orbits are identical in nature, but also that they are identical to a force familiar to everyone on the earth, the force that causes an apple to fall to the ground. In a word, the law of universal gravitation depends on the correlation of the centripetal acceleration of the moon with the acceleration of gravity on the surface of the earth—not an approximate correlation such as he had obtained in 1666, but an exact correlation.

Here another problem presented itself. As far as the sun and the planets are concerned, it appeared permissible to treat them as point masses, and even in the case of the moon and the earth, the bodies are not large in comparison to the distance separating them. The problem appears with the apple and the earth. At first blush, the apple on the tree appears to be ten or twenty feet away from the earth; the correlation Newton found demands that it be four thousand miles away. That is, Newton was using the distance of the apple from the center of the earth. (See Fig. 8.4.) Hence the crucial importance of a section found in Book One, in which Newton examined the attraction of bodies composed of attracting particles. He demonstrated that an homogeneous sphere (or a sphere composed of homogeneous shells), made up of particles attracting with a force that varies inversely as the square of the distance, attracts any body external to it with a force proportional to its quantity of matter (or mass) and inversely proportional to the square of the body's distance from its center. That is, such a sphere attracts as though its entire mass were concentrated at its central point. With this demonstration, and with the exact correlation of the moon's centripetal acceleration with the acceleration of gravity, Newton was ready to state

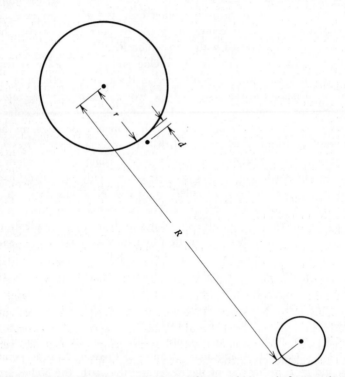

Figure 8.4. Earth, moon, and apple. The correlation of the centripetal acceleration of the moon, removed from the earth a distance R, with the centripetal acceleration of the apple demands that the distance of the apple from the earth be, not d, *but* r + d, *which is, for all practical purposes, equal to* r.

the law of universal gravitation: "That there is a power of gravity pertaining to all bodies, proportional to the several quantities of matter which they contain." The universe is composed of particles of matter all of which attract each other with a force proportional to the products of their masses and inversely proportional to the square of the distance between them.

Having derived the law of universal gravitation from the dynamic necessities inherent in the solar system, Newton then employed it in the rest of Book Three to explain a number of more complicated phenomena. During Newton's life, it was established that the length of a pendulum with a period of one second is shorter at the equator than it is in Europe. Newton derived the phenomenon with quantitative exactness from the law of universal gravitation. Tides had long held the interest of the scientific world; Newton showed that the attractions of the sun and the moon cause them—a significant confirmation of the mutuality of gravitational attraction. Of all the celestial bodies then known, the moon moved with the most irregularities. By treating it as a body attracted both by the earth and by the sun, Newton was able to show that the law of universal gravitation probably accounts for its irregularities. The problem was highly complicated, and Newton's lunar theory remained imperfect. When astronomers did perfect it in the 18th century, the law of universal gravitation received a considerable confirmation. Newton had greater success with the precession of the equinoxes, the slow oscillation of the earth's axis, and his greatest success of all was the solution of cometary orbits. Until Newton, comets had seemed to defy attempts to reduce their motion to law; he demonstrated that their motion is governed by the same dynamic laws that govern the motions of the planets.

Although Newton had invented the calculus some twenty years before the *Principia* was written, and before Leibniz's independent invention of it, he did not present his great work in its terms. Geometry continued to be regarded as the language of science, and geometry he employed. He did utilize the notion of ultimate and nascent ratios which resemble differentials in some respects. (See Fig. 8.5.) Newton's concepts and conclusions could be translated readily into the language of the calculus, however, and his followers of the 18th century employed Leibniz's version of the calculus to extend the range of Newtonian mechanics.

Gravitational attraction as Newton conceived it differs from the forces between particles discussed in Query 31 of the *Opticks*. Those forces were believed to be, not universal, but specific, one type of matter

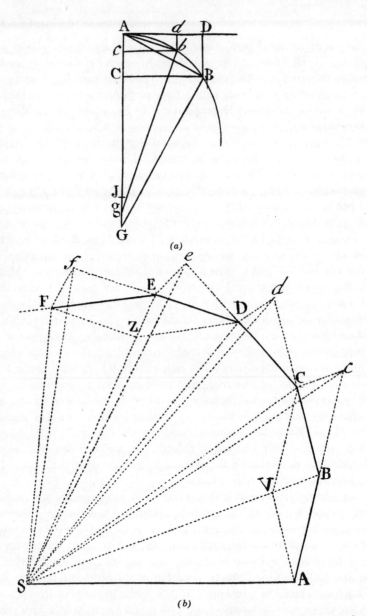

Figure 8.5. (a) *The ultimate ratio of* DB/AD *(or* db/Ad*) as* D *approaches* A *and the radius* GB *becomes coincident with the radius* JA *is a similar mathematical concept to that of the differential.* (b) *A typical diagram from the* Principia. *As the problem is defined, separate discrete centripetal impulses change the direction of the body at* A, B, C, D, E, *and* F. *The number of sides of the polygon is then allowed to increase, and the ratio of the constant centripetal force to the momentum at* B *is the ultimate ratio of* BV *to* AB *when the polygon approaches a smooth curve.*

acting only on other matter related to it, as the magnet attracts iron but not copper. Gravity, in contrast, was held to be an attraction by which all matter, as matter, attracts all other matter. It was universal, and as universal it affirmed a basic tenet of the mechanical philosophy, the identity of all matter. Nevertheless, mechanical philosophers became more than a little uneasy in the presence of suggested attractions. In 1687, when the *Principia* was nearing publication, a young Swiss mathematician who had come to England and met Newton, wrote to Huygens telling him of the coming work on the system of the world. Huygens replied that he looked forward to seeing it, but hoped that it would not be another theory of attraction. Alas, it was exactly that, and mechanical philosophers threw up their hands in dismay. What was gravitational attraction? Either it had a mechanical cause, which Newton should explain, or it was an occult quality, which was inadmissible.

In the end, Newton acknowledged the criticism to the extent of including a new set of eight Queries (numbers 17 through 24) in the second English edition of the *Opticks* in 1717, in which he explained the action of gravity by means of an aether pervading the universe. The compromise was only apparent, however, since the new aether was composed of particles which repelled each other at a distance. There can be no serious doubt that Newton considered the forces between particles as ontological realities, and not just appearances. In so far as he was prepared to discuss their cause at all, he referred them directly to the agency of God.

The question of what convinced Newton to admit a new category to the ontology of the mechanical universe remains, however, and the answer must be given primarily in terms of his ideal of science. An exchange early in the 1690s between Huygens and Leibniz, discussing Newton's theory, throws some light on their difference from Newton. The planets do not merely travel in ellipses, Leibniz said, but all of them move in the same plane in the same direction around the sun. By Newton's theory they should be able to move in any plane in any direction; hence some resort to the vortex, which does explain these things, is necessary. Leibniz's letters repeated the basic conviction of the mechanical philosophy, that the universe is transparent to human reason. In contrast, the law of universal gravitation appeared to impose an opaque screen at one level of understanding. Fifteen years before the *Principia,* Huygens' response to Newton's paper on color had made the same criticism—Newton had merely shown that the rays which produce the sensation of one color are differently refrangible from those which produce another, but he had not explained what the differences in color are. The basic issue between Newton and the traditional mechanical philosophers lay in his willingness to employ an ideal of science which accepted the ultimate inscrutability of nature.

The concept of force brought the central issues of his ideal of science to a focus. In Query 31, originally published in 1706, Newton replied to the charge that he was reinstating occult qualities.

"These Principles I consider, not as occult Qualities, supposed to result from the specifick Forms of Things, but as general Laws of Nature, by which the Things themselves are formed; their Truth appearing to us by Phaenomena, though their Causes be not yet discovered. For these are manifest Qualities, and their Causes only are occult. To tell us that every Species of Things is endowed with an occult specifick Quality by which it acts and produces manifest Effects, is to tell us nothing: But to derive two or three general Principles of Motion from Phaenomena, and afterwards to tell us how the Properties and Actions of all corporeal Things follow from those manifest Principles, would be a very great step in Philosophy, though the Causes of those Principles were not yet discovered: And therefore I scruple not to propose the Principles of Motion above-mentioned, they being of very general Extent, and leave their Causes to be found out."

In the General Scholium published in 1713 at the end of the second edition of the *Principia,* Newton expressed the same point of view in its classic form. So far, he said, he had explained the phenomena by means of gravity, but he had not explained the cause of gravity,

"But hitherto I have not been able to discover the cause of those properties of gravity from phaenomena, and I feign no hypotheses; for whatever is not deduced from the phaenomena is to be called an hypothesis; and hypotheses, whether metaphysical or physical, whether of occult qualities or mechanical, have no place in experimental philosophy."

I feign no hypotheses—*hypotheses non fingo.* In one sense the words are obviously false; Newton did feign hypotheses, and rather grandiose ones at that. In the sense that he maintained a rigid distinction between demonstrated conclusions and hypotheses meant to explain them, and refused to dilute demonstrations with speculations, however, the statement can stand. Thus force was to Newton a concept necessary to the description of phenomena in mechanical terms. Its validity rested on its utility in demonstrations, not on hypotheses that might explain its origin.

Newton believed that nature is ultimately opaque to human understanding. Science cannot hope to obtain certain knowledge about the essences of things. Such had been the program of the mechanical philosophy in the 17th century, and the constant urge to imagine invisible mechanisms sprang from the conviction that a scientific explanation is only valid when it traces phenomena to ultimate entities. To Newton, in contrast, nature was a given, aspects of which might never be intelligible. When they learned to accept the same limitation, other sciences such as optics, chemistry, and biology, likewise ceased to play with imaginary mechanisms and, describing instead of explaining, they formulated a set of conceptions adequate to their phenomena. Newton believed that the aim of physics is an exact description of phenomena of motion in quantitative terms. Thus the concept of force could be admitted into scientific demonstrations even if the ultimate reality of force were not comprehended. In Newton's work, it made possible the reconciliation of the tradition of mathematical description, represented by Galileo, with the tradition of mechanical philosophy, represented by Descartes. By uniting the two, Newton carried the scientific work of the 17th century to that plane of achievement which has led historians to speak of a scientific revolution. And modern science continues to pursue its effective course within the framework thus established.

Suggestions for Further Reading

The works of Alexandre Koyré have done more than any other one factor to shape the present understanding of science in the 17th century. Unfortunately, his basic work, *Etudes Galiléennes,* (Paris, 1939) has not been translated into English, but shorter articles that summarize its conclusions have been reprinted in a volume entitled *Metaphysics and Measurement* (Cambridge, Mass., 1968). Another book, *From the Closed World to the Infinite Universe* (Baltimore, 1957) also presents Koyré's views on the period. Scarcely less important for comprehending the basic intellectual currents is E. A. Burtt, *The Metaphysical Foundations of Modern Physical Science,* revised ed. (London, 1932). There are several excellent books which are readily available that concern themselves primarily with science in the 17th century. Herbert Butterfield, *The Origins of Modern Science,* (London, 1950), and two books by A. R. Hall, *The Scientific Revolution,* (London, 1954), and *From Galileo to Newton,* (New York, 1963), a more recent and chronologically more restricted study, are the most prominent of these. There are also valuable insights in the sections relative to the 17th century in two works which cover more extensive periods. E. J. Dijksterhuis, *The Mechanization of the World Picture,* trans. C. Dikshoorn, (Oxford, 1961), concludes with the 17th century; C. C. Gillispie, *The Edge of Objectivity,* (Princeton, 1960), begins with it. Both works have deservedly exercised a considerable influence. Finally, volume two of *History of Science,* ed. René Taton, trans. A. J. Pomerans (New York, 1964), which contains detailed articles on developments in the individual sciences, is a useful reference work.

The developments in astronomical thought from Copernicus through Kepler have been the subject of a considerable volume of literature.

Again, Koyré has contributed a major study, *La révolution astronomique,* (Paris, 1961) which is not available in English, but Max Casper's biography, entitled simply *Kepler,* trans. C. Doris Hellman, (New York, 1959) has been. Arthur Koestler, *The Sleepwalkers,* (London, 1959), a tendentious book if ever there was one, contains as its central feature an interesting portrait and analysis of Kepler, whom Koestler admires to the detriment of everyone else who appears in the book. W. Pauli, *The Influence of Archetypal Ideas on the Scientific Theories of Kepler* (New York, 1955), and Gerald Holton, "Johannes Kepler's Universe: Its Physics and Metaphysics," *American Journal of Physics,* 24 (1956), 340–351, contain important analyses of Kepler's thought. A reliable resume of his contribution to planetary theory is found in J. L. E. Dreyer, *History of the Planetary Systems from Thales to Kepler,* (Cambridge, 1906).

Koyré's fundamental contribution to the understanding of Galileo's work in mechanics has been mentioned above. He has also written a number of articles on mechanics during the 17th century, among which I shall cite only two—*A Documentary History of the Problem of Fall from Kepler to Newton, (Transactions of the American Philosophical Society,* New series, Vol. 45, Part 4, 1955), and "An Experiment in Measurement," *Proceedings of the American Philosophical Society,* 97 (1953), 222–237. A collective volume of articles on every aspect of Galileo's career, *Galileo, Man of Science,* ed. E. McMullin (New York, 1967), contains a number concerned with his mechanics, among which the title piece, by the editor McMullin, is of especial importance. Among general histories of mechanics during the century, I. B. Cohen, *The Birth of a New Physics* (Garden City, New York, 1960), contains a lucid exposition. Ernst Mach, *The Science of Mechanics: A Critical and Historical Account of its Development,* trans. T. J. McCormack, 6th ed. (LaSalle, Illinois, 1960), is a classic of critical analysis which is basically historical in its organization. For those with real endurance, René Dugas, *Mechanics in the Seventeenth Century,* trans. F. Jacquot (Neuchatel, 1958), contains a great deal of information in a fairly indigestible form.

There has been a great deal of interest in the Hermetic tradition of late, most of it concentrating on figures before the 17th century, such as Paracelsus and Bruno. Paolo Rossi, *Francis Bacon: From Magic to Science,* trans. S. Rabinovitch (London, 1968), and Walter Pagel, *The Religious and Philosophical Aspects of van Helmont's Science and Medicine (Supplements to the Bulletin of the History of Medicine,* No.

2, Baltimore, 1944), are two notable exceptions that get into 17th century material. On the mechanical philosophy, there have been numerous studies. R. G. Collingwood, *The Idea of Nature* (Oxford, 1945), devotes one chapter to a perceptive analysis of the 17th century's conception of nature. Marie Boas (now Marie Boas Hall), "The Establishment of the Mechanical Philosophy," *Osiris,* 10 (1952), 412–541, studies just what its title suggests, concentrating its attention on Robert Boyle. R. Harré, *Matter and Method,* (London, 1964), presents an historically organized philosophical analysis. The vast bulk of material on Descartes has not devoted extensive consideration to Descartes as part of the scientific tradition, but most of the books on Cartesian philosophy have something about his conception of matter and nature. Gassendi and the atomist tradition have been studied much less and mostly in France, but there is a recent book by Robert Kargon, *Atomism in England from Hariot to Newton* (Oxford, 1966).

W. E. K. Middleton has recently published *The History of the Barometer* (Baltimore, 1964). Optics has been studied less extensively than one might expect, but an outstanding work by A. I. Sabra, *Theories of Light from Descartes to Newton* (London, 1967), which does not attempt to be a history of topics in the 17th century, comes closer to it than anything else. Vasco Ronchi, *Histoire de la lumière,* trans. J. Taton (Paris, 1956), is the standard survey of optics in general, including the 17th century. I have published several articles that explore more fully topics brought up here—"The Development of Newton's Theory of Colors," *Isis,* 53 (1962), 339–358; "Isaac Newton's Coloured Circles Twixt Two Contiguous Glasses," *Archive for History of Exact Sciences,* 2 (1965), 181–196; "Uneasily Fitful Reflections on Fits of Easy Transmission," *The Texas Quarterly,* 10 (1967), 86–102; and "Hugyens' Rings and Newton's Rings: Periodicity and 17th Century Optics," *Ratio,* 10 (1968), 64–77.

The most important histories of biological thought in the 17th century are in French—Emile Guyenot, *Les sciences de la vie aux XVIIe et XVIIIe siècles* (Paris, 1941), and Jacques Roger, *Les sciences de la vie dans la pensée francaise du XVIIIe siècle* (Paris, 1963), which, despite the title, discusses the 17th century extensively. Erik Nordenskiold, *The History of Biology,* trans. L. B. Eyre (New York, 1935), the standard history of biology, contains a great deal of material on the 17th century. Harvey has been studied extensively. There are numerous biographies, among which are Robert Willis, "The Life of William Harvey, M.D.," in *The Works of William Harvey, M.D.* (London,

1848), and Geoffrey Keynes, *The Life of William Harvey* (Oxford, 1966). More specialized studies of his work on the heart are found in Charles Singer, *The Discovery of the Circulation of the Blood* (London, 1922), and H. P. Bayon, "William Harvey, Physician and Biologist," *Annals of Science*, 3 (1938), 59–118, 435–456; 4 (1939), 65–106, 329–389. The recent work by Walter Pagel, *William Harvey's Biological Ideas* (Basel & New York, 1967), a masterpiece by an outstanding historian of science, sets his work in the context of his whole approach to biology. Embryology in the 17th century is only beginning to be studied in detail, but there is a recent massive publication of both sources and secondary discussion of them by Howard Adelmann, *Marcello Malpighi and the Evolution of Embryology*, 5 vols. (Ithaca, New York, 1966).

The basic book on chemistry in the 17th century also remains untranslated from the French—Hélène Metzger, *Les doctrines chimiques en France du début du XVIIe à la fin du XVIIIe siècle* (Paris, 1923). Miss Metzger did not live to realize the entire program indicated in the title, but she covered chemistry in the 17th century both in the volume above and in another entitled *Newton, Stahl, Boerhaave et la doctrine chimique* (Paris, 1930). There are two fine works which together cover English chemistry in the 17th century—Allen Debus, *The English Paracelsians* (London, 1965), and Marie Boas (Hall), *Robert Boyle and Seventeenth-Century Chemistry* (Cambridge, 1958). Thomas Kuhn, "Robert Boyle and Structural Chemistry in the Seventeenth Century," *Isis,* 43 (1952), 12–36, sheds more light on chemistry in the age of the scientific revolution than any other twenty-five pages I know of.

The development of the scientific societies in the 17th century has been the subject of a number of studies. Martha Ornstein, *The Role of the Scientific Societies in the Seventeenth Century* (Chicago, 1928) is the standard general history. The Royal Society has attracted continuing study. The most recent histories of it are Dorothy Stimson, *Scientists and Amateurs, A History of the Royal Society* (New York, 1948), and Margery Purver, *The Royal Society: Concept and Creation* (London, 1967), the latter being a somewhat argumentative book which attempts to impose a Baconian pattern on the organization. Harcourt Brown, *Scientific Organizations in Seventeenth Century France* (Baltimore, 1934), presents detailed accounts of the French societies to the reader of English. Two recent books present somewhat different views of science in the universities—William T. Costello, *The Scholastic Cur-*

riculum at Early Seventeenth-Century Cambridge (Cambridge, Mass., 1958), and Mark Curtis, *Oxford and Cambridge in Transition 1558–1642* (Oxford, 1959). R. K. Merton, "Science, Technology and Society in Seventeenth Century England," *Osiris,* 4 (1938), 360–632, and Edgar Zilsel, "The Sociological Roots of Science," *American Journal of Sociology,* 47 (1941–42), 544–562, are pioneering studies of the social background of the scientific revolution. A. R. Hall has taken issue with their conclusions in two articles—"Merton Revisited," *History of Science,* 2 (1963), 1–16, and "The Scholar and the Craftsman in the Scientific Revolution," *Critical Problems in the History of Science,* ed. Marshall Clagett (Madison, 1962), pp. 3–23. There is no history of the development of scientific method; R. M. Blake, C. J. Ducasse, and E. H. Madden, *Theories of Scientific Method* (Seattle, 1960), is the nearest approach. R. F. Jones, *Ancients and Moderns* (St. Louis, 1936), is an important study of various attitudes associated with the scientific movement.

On Isaac Newton there has been an outpouring of scholarly endeavor too immense even to be indicated here. There have been a number of biographies, of which the most important are David Brewster, *Memoirs of the Life, Writings, and Discoveries of Sir Isaac Newton* (Edinburgh, 1855), and Louis T. More, *Isaac Newton, A Biography* (New York, 1934). Frank Manuel, *A Portrait of Isaac Newton* (Cambridge, Mass., 1968), presents a brilliant historical psychoanalysis which everyone ought to read. As with every important topic in 17th century science, Koyré has written extensively on Newton; his most important articles have been collected in a volume entitled *Newtonian Studies* (Cambridge, Mass., 1965). Another fundamental study which cannot be neglected is I. B. Cohen, *Franklin and Newton* (Philadelphia, 1956). John Herivel, *The Background to Newton's "Principia"* (Oxford, 1965), presents a detailed analysis of the development of Newton's dynamics together with all the relevant sources. Another collection of papers edited by A. R. and Marie Boas Hall, *Unpublished Scientific Papers of Isaac Newton* (Cambridge, 1962), contains valuable introductions as does *Isaac Newton's Papers & Letters on Natural Philosophy,* ed. I. B. Cohen (Cambridge, Mass., 1958). The acknowledged master of Newton's mathematics is D. T. Whiteside ed., *The Mathematical Papers of Isaac Newton,* 3 vols. continuing (Cambridge, 1967 continuing). Whiteside's "Patterns of Mathematical Thought in the later Seventeenth Century," *Archive for History of Exact Sciences,* 1 (1961), 179–388, together with his introductions in the edition of the mathematical papers, is the best work on Newton's mathematics. A recent issue of the *Texas Quarterly* (Vol. 10, No. 3, 1967), which bears the general title "The *Annus Mirabilis* of Sir Isaac Newton," is a collection of articles that present a coherent picture of the present status of understanding of Newton.

It is impossible to leave the subject of bibliography without pointing out the ready availability in English of most of the basic works of science in the 17th century. Of Kepler, it is true, only two books from *The Epitome of Copernican Astronomy* and one from *The Harmonies of the World* have been translated in Volume 16 of *Great Books of the Western World*, R. M. Hutchins ed. (Chicago, 1952). All of Galileo's works are in translation and most of them are now in print. Most of Descartes is in translation and in print. A translation-epitome of Gassendi appeared in the 17th century. Of the two great classics of late 17th century optics, Newton's *Opticks* was written originally in English and Huygens' *Treatise on Light* has been translated. Nothing else by Huygens has been. Boyle's works appeared originally in English for the most part, and Van Helmont's were put into English in the 17th century. Gilbert has been translated and is in print, and the same is true of Harvey. Recently Malpighi's work on embryology was translated together with Gassendi's. Hooke's classic microscopical observations, published in English despite their Latin title, have been republished periodically ever since and are now in print. Most of Leibniz's essays and works have been translated. All of Newton's major works are in print, the *Principia* in translation, and a considerable volume of his papers as well. The scientific revolution is more accessible in its original works than in any number of secondary accounts.

Index